Statistical Test Theory for the Behavioral Sciences

Chapman & Hall/CRC
Statistics in the Social and Behavioral Sciences Series

Series Editors

Andrew Gelman
Columbia University, USA

J. Scott Long
Indiana University, USA

Sophia Rabe-Hesketh
University of California, Berkeley, USA

Anders Skrondal
London School of Economics, UK

Aims and scope

Large and complex datasets are becoming prevalent in the social and behavioral sciences and statistical methods are crucial for the analysis and interpretation of such data. This series aims to capture new developments in statistical methodology with particular relevance to applications in the social and behavioral sciences. It seeks to promote appropriate use of statistical, econometric and psychometric methods in these applied sciences by publishing a broad range of reference works, textbooks and handbooks.

The scope of the series is wide, including applications of statistical methodology in sociology, psychology, economics, education, marketing research, political science, criminology, public policy, demography, survey methodology and official statistics. The titles included in the series are designed to appeal to applied statisticians, as well as students, researchers and practitioners from the above disciplines. The inclusion of real examples and case studies is therefore essential.

Proposals for the series should be submitted directly to:

Chapman & Hall/CRC
Taylor and Francis Group
Informa
24-25 Blades Court
Deodar Road
London SW15 2NU, UK

Statistics in the Social and Behavioral Sciences Series

Statistical Test Theory for the Behavioral Sciences

Dato N. M. de Gruijter
Leo J. Th. van der Kamp

CRC Press
Taylor & Francis Group
Boca Raton London New York

CRC Press is an imprint of the
Taylor & Francis Group, an **informa** business
A CHAPMAN & HALL BOOK

CRC Press
Taylor & Francis Group
6000 Broken Sound Parkway NW, Suite 300
Boca Raton, FL 33487-2742

First issued in paperback 2019

© 2008 by Taylor & Francis Group, LLC
CRC Press is an imprint of Taylor & Francis Group, an Informa business

No claim to original U.S. Government works

ISBN-13: 978-1-58488-958-8 (hbk)
ISBN-13: 978-0-367-38867-6 (pbk)

Library of Congress Cataloging-in-Publication Data

Gruijter, Dato N. de.
    Statistical test theory for the behavioral sciences / Dato N.M. de Gruijter and Leo J. Th. van der Kamp.
        p. cm. -- (Statistics in the social and behavioral sciences series ; 2)
    Includes bibliographical references and index.
    ISBN-13: 978-1-58488-958-8 (alk. paper)
    1. Social sciences--Mathematical models. 2. Social sciences--Statistical methods. 3. Psychometrics. 4. Psychological tests. 5. Educational tests and measurements. I. Kamp, Leo J. Th. van der. II. Title.

H61.25.G78 2008
519.5--dc22   2007017631

Visit the Taylor & Francis Web site at
http://www.taylorandfrancis.com

and the CRC Press Web site at
http://www.crcpress.com

Table of Contents

Preface

What would science be without measurement, and what would social science be without social measurement? Social measurement belongs to the widely accepted and fruitful stream of the empirical-analytic approach. Statistics and research methodology play a central role, and it is difficult to ascertain precisely when it all started in the social sciences. An important subfield of social measurement is the quantification of human behavior—that is, using measurement instruments of which educational and psychological tests are the most prominent representatives. The intelligence test, for example, was developed in the early 20th century in France thanks to the research in school settings by Alfred Binet and Henri Simon. Actually, they were pioneers in social measurement at large. What applies to psychological and educational tests, also applies to social measurement procedures at large: measurement instruments must be valid and reliable in the first place. Many requirements of tests in education and psychology are also essential for social measurement.

We will thoroughly discuss the concepts of *reliability* and *validity*. In classical test theory, the test score is a combination of a true score and measurement error. It is possible to define the measurement error in several ways depending on the way one would like to generalize to other testing situations. Generalizability theory, developed from 1963 onward by Cronbach and his coworkers, effectively deals with this problem. It gives a framework in which the various aspects of test scores can be dealt with. Of much importance to test theory has been the development of item response theory, or IRT for short. In an item response model, or IRT model, the item is the unit of analysis instead of the test. In IRT models, the variance of measurement errors is a function of the level or ability of the respondent, an important characteristic that in most classical test theory models is not available in a natural way. IRT has resulted in improvements in test theoretical applications and in new applications as well, for example, in *computerized adaptive testing,* CAT for short.

This manuscript has been written for advanced undergraduate and graduate students in psychology, education, and other behavioral sciences. The prerequisites are a working knowledge of statistics including the basic concepts of the analysis of variance and regression analysis and some knowledge of estimation theory and methods. Of course, the more background in research methodology and statistical data analysis the reader has, the more he or she can profit. This text is also meant for researchers in the field of measurement and testing, not typically specialized in test theory. It portends not merely a broad overview but also a critical survey with hopefully knowledgeable comments and criticism on the test theories. An attempt is made to follow recent developments in the field. As aids in instruction, studying, and reading, each chapter concludes with exercises, the answers of which are given at the end of the book. Examples and exhibits are also included where they seemed useful.

There are some great books on mental test theory. Gulliksen (1950) and Lord and Novick (1968) should be mentioned first and with great deference. These are the godfathers of classical test theory, and they were the ones to codify it. Would generalizability theory have been developed without the work of Lee J. Cronbach (see, e.g., Cronbach Gleser, Nanda, and Rajaratnam, 1972). As in many fields of science, inventions and developments are not one man's achievement. So it is with item response theory, and therefore, being aware of doing injustice to other authors, we mention only Rasch (1960), Birnbaum (1968), Lord and Novick (1968), and Lord (1980). The *Standards for Educational and Psychological Testing* (American Psychological Association [APA], American Educational Research Association [AERA], and the National Council on Measurement in Education [NCME]) served as guidelines, and ample reference is made to them. For a more in-depth treatment of the psychometrical topics in this book, the reader is referred to volume 26 of the *Handbook of Statistics* (2007), *Psychometrics*, edited by Rao and Sinharay.

Information on test theory can readily be obtained from the World Wide Web. *Wikipedia* is one source of information. There certainly are other useful sites, but it is not always clear whether they remain available and if the presented information is of good quality. We decided to refer to only a few sites for software.

Previous versions of this book have been used in one-semester courses in test theory for advanced undergraduate and graduate students of psychology and education. Comments from our students were helpful in improving the text.

Dato N. M. de Gruijter

Leo J. Th. Van der Kamp

The Authors

Dato N. M. de Gruijter currently is senior advisor at the Graduate School of Teaching at Leiden University. He also teaches classes on test theory at the department of psychology at Leiden University. He received his Ph.D. in the social sciences from Leiden University. His principal interest is educational measurement. He published on the topics of generalizability theory and item response theory.

Leo J. Th. van der Kamp is emeritus professor of psychology at Leiden University. His research interests include research methodology, psychological test theory, and multivariate analysis. His current research is on quasi-experimental research and he is a perennial student of early Taoism. His publications are in the area of generalizability theory, item response theory, the application of multilevel modeling and structural equation modeling in health psychology and education, clinical epidemiology, and longitudinal data analysis for the social and behavioral sciences. He has taught many undergraduate and postgraduate courses on these topics and supervised more than 50 doctoral dissertations.

CHAPTER 1

Measurement and Scaling

1.1 Introduction

In behavioral sciences in general, and in education and psychology in particular, the use of measurement procedures or tests and assessments is ubiquitous. Measurement instruments are used for all kinds of assessments. The main types of psychological and educational tests are intelligence tests, aptitude tests, achievement tests, personality tests, interest inventories, behavioral procedures, and neuropsychological tests. The use of such tests is not restricted to psychology and education but stretches over other disciplines of the behavioral sciences, and even beyond (e.g., in the field of psychiatry). Using tests involves some kind of measurement procedure and, in addition, statistical theories for characterizing the results of the measurement procedures—that is, for modeling test scores.

In this chapter we will first give a broad and generally accepted definition of a test. Then a sketchy introduction will be given into measurement and scaling. Measurement not only pervades daily life, it is also the cornerstone of scientific inquiry. After defining the concept of measurement, scales of measurement and the relation between measurement and statistics will be presented. Some remarks will be made on scales of measurement in relation to the test theory models given later, while the concept of dimensionality of tests will also be discussed.

1.2 Definition of a test

A test is best defined as a standardized procedure for sampling behavior and describing it with categories or scores. Essentially, this definition includes systematic measurement in all fields of the behavioral sciences. This broad definition includes also checklists, rating scales, and observation schemes. The essential features of a test are that it is as follows:

- A standardized procedure, which means that the procedure is administered uniformly over a group of persons.
- A focused behavioral sample, which means that the test is focused on a well-defined behavioral domain. Examples of domains in educational measurement are achievement in arithmetic, or language performance. Psychological tests may also be targeted to constructs or theoretical variables (e.g., depression, extraversion, quality of life, emotionality, and the like), so, at variables that are not directly observable. In other words, such a measurement approach assumes that there exists a psychological attribute to measure. Such a psychological attribute is usually a core element of a *nomological network*, which maps its relations with other constructs, and also clarifies its relations with observables (i.e., relevant behavior in the empirical world).
- A description in terms of scores or mapping into categories. Using tests implies a form of measurement whereby performances, characteristics, and traits are represented in terms of numbers or classifications.

In addition to these features, once a test score is obtained, norms or standards of a relevant group of persons are necessary for the interpretation of the score of a given person. Finally, collecting test scores is seldom an aim in itself, the function of testing is ultimately decision making in a narrow as well as in a broad sense. This includes classification, selection and placement, diagnosis and treatment planning, self-knowledge, program evaluation, and research.

1.3 Measurement and scaling

Stevens defined measurement as "the assignment of numbers to aspects of objects or events according to one or another rule or convention" (Stevens, 1968, p. 850). Other, sometimes broader, sometimes more refined and more sophisticated definitions are around, but for our purpose Stevens' definition suffices. In addition to what is called psychometric measurement, considered here, representational measurement has been formulated. More can be found in Judd and McClelland (1998) and the references mentioned by them, or in Michell

(1999, 2005), who provides a critical history of the concept, and in McDonald (1999), who discusses measurement and scaling theory in the context of a unified treatment of test theory.

Usually a test consists of a number of items. The simplest item type is when only two answers are possible (e.g., *Yes* or *No, correct* or *incorrect*).

After a test has been administered to a group of persons, we generally have a score for each person. The simplest example of a test score is the total score on a multiple-choice test, where one point is given for a correct answer to an item and zero points are given for an incorrect answer or skipped item. Some persons have higher scores than others, and we expect that these differences are relevant.

We speak of a measurement once a score has been computed. The measurement refers to a property or aspect of the person tested. A well-known classification of measurement scales is given by Stevens (1951). These measurement scales are as follows:

1. The nominal scale—On the nominal scale, objects are classified according to a characteristic (e.g., a person can be classified with respect to sex, hair color, etc.).
2. The ordinal scale—On the ordinal scale, objects are ordered according to a certain characteristic (e.g., the Beaufort scale of wind force).
3. The interval scale—On the interval scale, equal scale differences imply equal differences in the relevant property. (For example, the Celsius and Fahrenheit scales for temperature are interval scales; a difference of 1° at the freezing point is as large as a difference of 1° at the boiling point of water.)
4. The ratio scale—The ratio scale has a natural origin as well as equal intervals. Length in meters and weight in kilograms are defined on a ratio scale, as is temperature on the Kelvin scale. Ratio scales are relatively rare in psychology because of the difficulty of defining a zero point. Can a person have zero intelligence?

Most researchers do not regard the use of the nominal scale as measurement. One should at least be able to make a statement about the amount of the property in question. Many researchers use an even narrower definition of measurement: they restrict themselves to scales that at least have interval properties.

With interval measurements of temperature, two scales are in use: the Celsius scale and the Fahrenheit scale. The scales are related to each other through a linear transformation: $°F = (9/5)°C + 32$. The linear transformation is a permissible transformation. With a linear transformation, the interval properties of the scale are maintained. When we have a ratio scale, a general linear transformation is not permissible while such a transformation effects a change of the origin (0). With a ratio scale, only multiplication with a constant is permitted. For example, one can measure length in centimeters instead of meters. With an ordinal scale, all monotonously increasing transformations are permitted.

The scale properties are relevant when one wants to compute measures characterizing distributions and apply statistical tests. When an ordinal scale is used, one generally is not interested in the average score. The median seems more appropriate and useful. On the other hand, statistics seldom is interested in the measurement level of a variable (Anderson, 1961). When a statistical test is used, it is important to know whether the distributional assumptions hold. Even if the assumptions are not fully met, statistical tests may be used if they are robust against violations of the assumptions.

The interpretation of the outcome of a statistical test, however, depends on the assumption with respect to the measurement level (Lord, 1954). And, as in some cases a nonlinear transformation might reverse the order of two means, we should decide which kind of transformations we are prepared to apply and which kind of transformations we judge as too extreme to be relevant. More on measurement scales and statistics is presented in Exhibit 1.1.

Exhibit 1.1   On measurement scales or "what to do with football numbers"

How devoted must a researcher be to Stevens' measurement-directed position? Is it permitted to calculate means and standard deviations on scores on an ordinal scale? Lord (1953) relates a story about a professor who retired early because of feelings of guilt for calculating means and standard deviations of test scores. The university gave this professor the concession for selling cloth with numbers for football players, and a vending machine, to assign numbers randomly. The team of freshmen football players protested after a while, because the numbers given to them were too low. The professor consulted a statistician. What should be done in the dispute with the complaining members of the freshman football team? Are their football numbers indeed too low? The daring and realistic statistician, without any hesitation whatsoever, turned to compute all kinds of measures, including means and standard deviations of football numbers. The professor protested that these football numbers did not even constitute an ordinal scale. The statistician, however, retorted: "The numbers don't know that. Since the numbers don't remember where they come from, they always behave just the same way regardless" (Lord, 1953, p. 751). The statistician concluded that it was highly implausible that the numbers of the team were a random sample. Needless to say, Lord's professor turned out to be convinced and lost his feelings of guilt. He even took up his old position.

Lord's narrative is basic to the so-called measurement-independent position. However, "the utmost care must be exercised in interpreting the results of arithmetic operations upon nominal and ordinal numbers; nevertheless, in certain cases such results are capable of being rigorously and usefully interpreted, at least for the purpose of testing a null hypothesis" (Lord, 1954, p. 265).

In practice we may generally assume that the score scales of psychological and educational tests are not interval scales. Nevertheless, researchers frequently act as if the score scale is an interval scale. One might say that no harm is done as long as the predictions from this way of interpreting test results are useful. When difference scores are used as an indication of a learning result or an improvement and these scores are related to other variables, certainly the interval property is invoked. In other test theoretical applications, for example in nonlinear equating of tests—here tests differing in difficulty level and other scale aspects are scaled to the same scale—the interval property is implicitly rejected. In item response models, scores on different tests are nonlinearly related to each other. With these models, scores can be computed on a latent scale, and within the context of a particular model, the scale has the interval property. The remaining question is whether this interval property is a fundamental property of the characteristic or just a property that is a consequence of the scale representation chosen. The Rasch model, for example, has two representations of the characteristic measured: one representation on an additive scale (which is a special case of the interval scale) and another representation with a multiplicative model.

In many applications it is assumed that one dimension underlies the responses to the items of the test in question (see Exhibit 1.2). In principle, in intelligence testing, for example, various abilities interplay in the process of responding to the test item. Take the following as an example. In order to be able to respond correctly to mathematics items, the persons or examinees in the target population must be able to read the test instructions. Reading ability is needed, but it can be ignored because it does not play a role in the differences between persons tested. Some authors, however, argue that responses are always determined by more than one factor. In ability testing factors like speed, accuracy, and continuance have a role (Furneaux, 1960; Wang and Zhang, 2006; Wilhelm and Schulze, 2002).

Exhibit 1.2   Dimensionality of tests and items

Once measurement became common practice in scientific research in the behavioral sciences, the concept of dimensionality, or more specifically the concept of unidimensionality, emerged as a crucial requirement for measurement.

Two early psychometricians, Thurstone and Guttman, already stressed the importance of unidimensionality for constructing good measures, without using the term though:

"The measurement of any object or entity describes only one attribute of the object measured. This is a universal characteristic of all measurement" (Thurstone, 1931, p. 257).

"We shall call a set of items of common content a scale if (and only if) a person with a higher rank than another person is just as high or higher on every item than the other person" (Guttman, 1950, p. 62).

Definitions of dimensionality abound. Gessaroli and De Champlain (2005) focus their attention on definitions based on the principle of local independence, a principle that will be discussed more extensively within the context of item response models. Gessaroli and De Champlain describe methods to assess dimensionality and also list relevant software packages.

In classical test theory, no explicit assumption is made with respect to the dimensionality of tests. Some tests are useful just because the items are not restricted to a small domain of unidimensional items but belong to a broader, more articulated domain of interest. In generalizability theory, the possibility to generalize to a heterogeneous domain of reactions is explicitly present. In an anxiety questionnaire one might, for example, ask whether anxiety is raised in a number of different situations, and it is assumed that for respondents anxiety is partly situational. But if a researcher is interested in growth or change, test dimensionality is an important issue. For if the test responses are determined by more than one dimension, it is not clear which dimension is responsible for a change in the test responses.

Even when it can be deduced from test results that the test is unidimensional, one should not conclude that one trait or characteristic determines the responses. One should not mistakenly conclude from a consistency in responses that respondents actually possess a particular trait. When we speak here of abilities or (latent) traits, this is meant for the sake of succinctness; the responses can be described as if the respondents possess a certain latent trait.

In the one-dimensional item response models that will be discussed, the responses to the different test items are a measure for an underlying latent trait—that is, the expected score is an increasing function of the underlying trait. In this context the test items as well as the persons are positioned on the underlying trait or dimension. This is also called the scaling or mapping of items and persons on the same underlying dimension.

Exercises

1.1 Two researchers evaluate the same educational program. Researcher A uses an easy test as a pretest and posttest, researcher B uses a relatively difficult test. Is it likely that their results will differ? If that is the case, in which way are the results expected to differ?

1.2 In a tennis tournament, five persons play in all different combinations. Player A wins all games; B wins from C, D, and E; C wins from D and E; and D wins from E. The number of games won is taken as the total score. Which property has this score in terms of Stevens' classification?

CHAPTER 2

Classical Test Theory

2.1 Introduction

It is a trite observation that all human endeavors are replete with error. And the human endeavor of science is no exception. We err in our measurements—that is to say, how hard we may try, never will our measurements be perfect. "O heaven! Were man but constant, he were perfect. That one error fills him with faults; makes him run through all the sins. Inconstancy falls off ere it begins" (Shakespeare: The Two Gentlemen of Verona, Act v. iv. 110–114). Inconsistency is not the only error.

There are many possible ways to err in measurement. In other words, there are many sources of errors. These sources may vary depending on the particular branch of science involved. The question now is to tackle the problem of errors of measurement. The answer to this question appears to be simple—develop a theory of errors, or some would say, set up an error model. Indeed, this is an approach that has been followed for more than a century. And the earliest theory around is classical test theory.

Classical test theory is presented in this chapter. By defining true score, an explicit, abstract formulation of measurement error is given. This will be the theme of the next section. In Section 2.3 further details will be given on the population of subjects or persons, a topic relevant for further developing test theory, more specifically, for deriving reliability estimates. The central assumptions of classical test theory will also be given. These are relevant for reliability, and for considering various types of equivalence or comparability of test forms.

2.2 True score and measurement error

Suppose that we obtained a measurement x_{pi} on person p with measurement instrument i. Let us assume, for example, that we read the weight of this person from a particular weighing machine and registered the outcome. Next, we take a new measurement and we notice a difference from the first. The obtained measurements can be thought of as arising from a probability distribution for measurements X_p with realizations x_p.

With measurement in the behavioral sciences, we have a similar situation. We obtain a measurement and we expect to find another outcome from the measuring procedure if we would be able to repeat the procedure and replicate the measurement result. However, in the behavioral sciences we frequently are not able to obtain a series of comparable measurement results with the same measurement instrument because the measurements may have their impact on the person from whom measurements are taken. Memory effects prevent independent replications of the measurement procedure. We might, however, administer a second test constructed for measuring the same construct and notice that the person obtains a different score on this test than on the first test. So, here comes in the development of an appropriate theory of errors or error model. The simplest is the following. The underlying idea is that the observed test score is contaminated by a measurement error. The observed score is considered to be composed of a true score and a measurement error (see also Figure 2.1):

$$x_p = \tau_p + e_p \tag{2.1}$$

If the measurement could be repeated many times under the condition that the different measurements are experimentally independent, then the average of these measurements would give a reasonable approximation to τ_p. In formal terms, true score is defined as the expected value of the variable X_p (x_p from Equation 2.1 is a realization of the random variable X_p):

$$\tau_p = \mathrm{E}X_p \tag{2.2}$$

where E represents the expectation over independent replications.

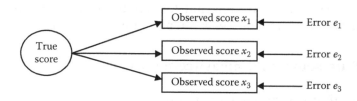

Figure 2.1 The decomposition of observed scores in classical test theory.

The definition of true score as an expected value seems obvious if the measurements to be taken can be considered exchangeable. In other words, this definition seems obvious if we do not know anything about a particular measurement. But consider the situation in which different measurement instruments are available and we have information on these instruments. For example, assume we have some raters as measurement instruments. Assume also that the raters differ in leniency, a fact known to occur. Does the definition of true score as an expected value do justice to this situation? Should we not correct the scores given by a rater with a known constant bias? The answer is that we can correct the scores without rejecting the idea of a true score, for it is possible to use the score scale of a particular rater and define a true score for this rater. Scores obtained on this scale can be transformed to another scale, comparable to the transformation of degrees Fahrenheit into degrees Celsius. The transformation of scores to scales defined by other measurement instruments will be discussed in Chapter 11.

In other situations, the characteristics of a particular rater are unknown. It is not necessary to have information on this rater, because the next measurement is likely to be taken by another rater. Then the rater effect can be considered part of the measurement error. In Exhibit 2.1, more information on multiple sources of measurement error is given.

The foregoing means that the definition of measurement error and, consequently, the definition of true score depend on the situation in which measurements are taken and used. If a particular aspect of the measurement situation has an effect on the measurements and if this aspect can be considered as fixed, one can define true score so as to incorporate this effect. This is the case when one tries to minimize noise in the data to be obtained through the testing procedure by standardization. In other cases, one is not able or not prepared to fix an aspect, and the variation due to fluctuations in the measurement context is considered part of the measurement error.

Exhibit 2.1   Measurement error: Systematic and unsystematic

Classical test theory assumes unsystematic measurement errors. Systematic measurement error may occur when a test consistently measures something other than the test purports to measure. A depression inventory, for example, may not merely tap depression as the intended trait to measure, but also anxiety. In this case, a reasonable decomposition of observed scores on the depression inventory would be

$$X = \tau + E_D + E_U$$

where X is the observed score, τ is the true score, E_D is the systematic error due to the anxiety component, and E_U is the combined effect of unsystematic error.

Clearly, the decomposition of observed score according to classical test theory is the most rudimentary form of linear model decomposition. Generalizability theory (see Chapter 5) has to say more on the decomposition of observed scores. Structural equation modeling might be used to unravel the components of observed scores.

Classical test theory can deal with only one true score and one measurement error. Therefore, the test researcher or test user must formulate precisely which aspects belong to the true score and which are due to measurement error. This choice also restricts the choice of methods to estimate reliability, which is the extent to which obtained score differences reflect true differences. Suppose we want to measure a characteristic that fluctuates from day to day, but which also is relatively stable in the long term. We might be interested in the momentary state, or in the expectation on the long term. If we are interested in measuring the momentary state, the value of the test–retest correlation does not have much relevance. A systematical framework for the many aspects of measurement errors and true scores was developed in generalizability theory.

From the definition of true score, we can deduce that the measurement error has an expected value equal to zero:

$$\mathrm{E}E_p = 0 \tag{2.3}$$

The variance of measurement errors equals

$$\sigma^2(E_p) = \sigma^2(X_p) \tag{2.4}$$

The square root from the variance in Equation 2.4 is the standard error of measurement for person p, the person-specific standard error of measurement.

2.3   The population of persons

To this point, we have treated measurements restricted to one person. In practice, we usually deal with groups of persons. If a person is tested, the test score is always interpreted within the context of measurements previously obtained from other persons. Test theory is concerned with measurements defined within a population or subpopulation of persons. An intelligence test, for example, is meant to be used for persons within a given age range, able to understand the test instructions. A population can be large or small.

Selecting a person randomly from the population, we have, analogous to Equation 2.1,

$$X = T + E \qquad (2.5)$$

where T (the Greek capital tau) designates the true-score random variable.

From the definitions given, the four central assumptions of classical test theory are as follows:

I The expected measurement error equals 0 (we take the expectation of the person-specific distribution of measurement errors over the population):

$$E_p E_p = 0 \qquad (2.6)$$

II The correlation ρ between measurement error and true score is 0 in the population:

$$\rho(T,E) = 0 \qquad (2.7)$$

This follows from the fact that the expected measurement error is equal (equal to 0) for all values τ.

We also assume that two measurements i and j are experimentally independent. From this assumption (actually from the weaker assumption of linearly experimental independence), we can deduce III and IV.

III For two measurements i and j holds that the true score on one measurement is uncorrelated with the measurement error on the second measurement:

$$\rho(T_i, E_j) = 0 \qquad (2.8)$$

IV Moreover, the measurement errors of the two measurements are uncorrelated:

$$\rho(E_i, E_j) = 0 \qquad (2.9)$$

For the population of persons, we can also deduce the equality of
the observed population mean and the true-score population mean:

$$EX = ET = \mu_X = \mu_T \tag{2.10}$$

The result in Equation 2.10 is obvious as well as important. In Equation 2.10, expectations are involved. The observed mean of a small
(sub)population certainly is not equal to the true-score mean. The
average measurement error may be small but is unlikely to be exactly
equal (0).

The variance of measurement errors can be written as

$$\sigma_E^2 = E_p \sigma^2(E_p)$$

and the variance of observed scores can be written as

$$\sigma_X^2 = \sigma_T^2 + \sigma_E^2 + 2\sigma_E \sigma_T \rho_{TE}$$

The correlation between true score and error is equal to zero, so
we can write the variance of observed scores as

$$\sigma_X^2 = \sigma_T^2 + \sigma_E^2 \tag{2.11}$$

The observed-score variance equals the sum of the variance of true
scores and the variance of measurement errors.

Exercises

2.1 A large testing agency administers test X to all candidates
at the same time in the morning. Other test centers organize
sessions at different moments. Give alternative definitions
of true score.

2.2 Two intelligence tests are administered close after one
another. What kind of problem do you expect?

CHAPTER 3

Classical Test Theory and Reliability

3.1 Introduction

Classical test theory gives the foundations of the basic true-score model, as discussed in Chapter 2. In this chapter, we will first go into some properties of the classical true-score model and define the basic concepts of reliability and standard error of measurement (Section 3.2). Then the concept of parallel tests will be discussed. Reliability estimation will be considered in the context of parallel tests (Section 3.3). Defining the reliability of measurement instruments is theoretically straightforward; estimating reliability, on the other hand, requires taking into account explicitly the major sources of error variance. In Chapter 4, the most important reliability estimation procedures will be discussed more extensively.

The reliability of tests is, among others, influenced by test length (i.e., the number of parts or items in the test) and by the homogeneity of the group of subjects to whom the test is administered. This is the subject of Sections 3.4 and 3.5. Section 3.6 is concerned with the estimation of subject's true scores. Finally, we could ask ourselves what the correlation between two variables X and Y would be "ideally" (i.e., when errors of measurement affect neither variable). In Section 3.7 the correction for attenuation is presented.

3.2 The definition of reliability and the standard error of measurement

An important development in the context of the classical true-score model is that of the concept of reliability. Starting from the variances and covariances of the components of the classical model, the concept of reliability can directly be defined. First, consider the covariance between observed scores and true scores. The covariance between observed and true scores, using the basic assumptions of the classical model discussed in Chapter 2, is as follows:

$$\sigma_{XT} = \sigma(T+E,T) = \sigma_T^2 + \sigma(T,E) = \sigma_T^2$$

Now the formula for the correlation between true scores and observed scores can be derived as

$$\rho_{XT} = \frac{\sigma_{XT}}{\sigma_X \sigma_T} = \frac{\sigma_T}{\sigma_X}$$

the quantity also known as the reliability index. The reliability of a test is defined as the squared correlation between true scores and observed scores, which is equal to the ratio of true-score variance to observed-score variance:

$$\rho_{XT}^2 = \frac{\sigma_T^2}{\sigma_X^2} = \frac{\sigma_T^2}{\sigma_T^2 + \sigma_E^2} \tag{3.1}$$

The reliability indicates to which extent observed-score differences reflect true-score differences. In many test applications, it is important to be able to discriminate between persons, and a high test reliability is prerequisite. A measurement instrument that is reliable in a particular population of persons is not necessarily reliable in another population. From Equation 3.1, it is clear that the size of the test reliability is population dependent. In a population with relatively small true-score differences, reliability is necessarily relatively low.

Estimation of test reliability has always been one of the important issues in test theory. We will discuss reliability estimation extensively in the next chapter. For the moment, we assume that reliability is known. Now we can define the concept of standard error of measurement. We derive the following from Equation 3.1:

$$\sigma_T^2 = \rho_{XT}^2 \sigma_X^2 \tag{3.2}$$

and

$$\sigma_E^2 = \sigma_X^2 - \rho_{XT}^2 \sigma_X^2$$

The standard error of measurement is defined as

$$\sigma_E = \sigma_X \sqrt{1 - \rho_{XT}^2}$$ (3.3)

The reliability coefficient of a test and the standard error of measurement are essential characteristics (cf. *Standards*, APA, AERA, and NCME, 1999, Chapter 2). From the theoretical definition of reliability (Equation 3.1), and taking into account that variances cannot be negative, the upper and lower limits of the reliability coefficient can easily be derived as

$$0 \le \rho_{XT}^2 \le 1.0$$

and $\rho_{XT}^2 = 0$ if all observed-score variance equals error variance. If no errors of measurement occur, observed-score variance is equal to true-score variance and the measurement instrument is perfectly reliable (assuming that there is true-score variation).

The observed-score variance is population or sample dependent, as is the reliability coefficient. Reporting only the reliability coefficient of a test is insufficient—the standard error of measurement must also be reported.

3.3 The definition of parallel tests

Generally speaking, parallel tests are completely interchangeable. They are perfectly equivalent. But how can equivalence be cast in statistical terms? Parallel tests are defined as tests that have identical true scores and identical person-specific error variances. Needless to say, parallel tests must measure the same construct or underlying trait.

For two parallel tests X and X', we have, as defined,

$$\tau_p = \tau_p' \quad \text{for all persons } p \text{ from the population}$$ (3.4a)

and

$$\sigma_{E_p}^2 = \sigma_{E_p'}^2 \quad \text{for all } p$$ (3.4b)

Using the definition of parallel tests and the assumptions of the classical true-score model, we can now derive typical properties of two parallel tests X and X':

$$\mu_X = \mu_{X'} \tag{3.5a}$$

$$\sigma_E^2 = \sigma_{E'}^2 \tag{3.5b}$$

$$\sigma_T^2 = \sigma_{T'}^2 \tag{3.5c}$$

$$\sigma_X^2 = \sigma_{X'}^2 \tag{3.5d}$$

and

$$\rho_{XY} = \rho_{X'Y} \quad \text{for all tests } Y \text{ different from tests } X \text{ and } X' \tag{3.5e}$$

In other words, strictly parallel tests have equal means of observed scores; equal observed-score, true-score, and error-score variances; and equal correlations with any other test Y.

Now working out the correlation between two parallel tests X and X', it follows that

$$\rho_{XX'} = \frac{\sigma_{TT'}}{\sigma_X \sigma_{X'}} = \frac{\sigma_T^2}{\sigma_X^2} = \rho_{XT}^2 \tag{3.6}$$

A second theoretical formulation of test reliability is that it is the correlation of a test with a parallel test. With this result, we obtained the first possibility to estimate test reliability: we can correlate the test with a parallel test. A critical note with this method, however, is how we should verify whether a second test is parallel. Also, parallelism is not a well-defined property: a test might have different sets of parallel tests (Guttman, 1953; see also Exhibit 3.1). Further, if we do not have a parallel test, we must find another way to estimate reliability.

Exhibit 3.1 On parallelism and other types of equivalence

To be sure, a certain test may have different sets of parallel tests (Guttman, 1953). Does it matter, for all practical purposes, if a test has different sets of parallel forms? An investigator will always look for meaningfulness and interpretability of the measurement results. If certain parallel forms do not suit the purpose of an investigator using a specific test, this investigator might well choose the most appropriate form of parallel test. Appropriateness may be checked against criteria relevant for the study at issue.

Parallel tests give rise to equal score means, equal observed-score and error means, and equal correlations with a third test. Gulliksen (1950) mentions the Votaw–Wilks' tests for this strict parallelism. These tests, among others, are also embedded in some computer programs for what is known as *confirmatory factor analysis*. "Among others" implies that other types of equivalence can also be tested statistically by confirmatory factor analysis.

3.4 Reliability and test length

In general, to obtain more precise measurements, more observations of the same kind have to be collected. If we want a precise measure of body weight, we could increase the number of observations. Instead of one measurement, we could take ten measurements, and take the mean of these observations. This mean is a more precise estimate of body weight than the result of a single measurement. This is what elementary statistics teaches us. If we have a measurement instrument for which two or more parallel tests are available, we might consider the possibility of combining them into one longer, more reliable test. Assume that we have k parallel tests. The variance of the true scores on the test lengthened by a factor k is

$$\mathrm{var}(k\mathrm{T}) = k^2\sigma_\mathrm{T}^2$$

Due to the fact that the errors are uncorrelated, the variance of the measurement errors of the lengthened test is

$$\mathrm{var}(E_1 + E_2 + \ \ + E_k) = k\sigma_E^2$$

The variance of the measurement errors has a lower growth rate than the variance of true scores.

The reliability of the test lengthened by a factor k is

$$\rho_{X(k)X'(k)} = \frac{k^2\sigma_\mathrm{T}^2}{k^2\sigma_\mathrm{T}^2 + k\sigma_E^2} = \frac{k\sigma_\mathrm{T}^2}{k\sigma_\mathrm{T}^2 + \sigma_X^2 - \sigma_\mathrm{T}^2}$$

After dividing numerator and denominator of the right-hand side by σ_X^2, we obtain

$$\rho_{X(k)X'(k)} = \frac{k\rho_{XX'}}{1+(k-1)\rho_{XX'}} \tag{3.7}$$

This is known as the general *Spearman–Brown formula for the reliability of a lengthened test.*

It should be noted that, as mentioned earlier, the lengthened test must measure the same characteristic, attribute, or trait as the initial test. That is to say, some form of parallelism is required of the supplemented parts with the initial test. Adding a less-discriminating item might lower test reliability. For (partly) speeded tests, adding items to boost reliability has its specific problems. Lengthening a partly speeded multiple-choice test might also result in a lower reliability (Attali, 2005).

3.5 Reliability and group homogeneity

A reliability coefficient depends also on the variation of the true scores among subjects. So, the homogeneity of the group of subjects is an important characteristic to consider in the context of reliability. If a test has been developed to measure reading skill, then the true scores for a group of subjects consisting of children of a primary school will have a wider range, or a larger true-score variance, than the true scores of, for example, the fifth-grade children only. If we assume, as is frequently done, that the error-score variance is equal for all relevant groups of subjects, we can compute the reliability coefficient for a target group from the reliability in the original group:

$$\rho_{UU'} = 1 - \frac{\sigma_E^2}{\sigma_U^2} = 1 - \frac{\sigma_X^2(1-\rho_{XX'})}{\sigma_U^2} \tag{3.8}$$

where σ_U^2 is the variance of the observed scores in the target group, σ_X^2 its counterpart in the original group, and $\rho_{XX'}$ the reliability in the original group.

It is, however, advised to verify whether the size of the error variance varies systematically with the true-score level. One method for the computation of the conditional error variance, an important issue for reporting errors of measurement of test scores (see *Standards*, APA et al., 1999, Chapter 2) has been suggested by Woodruff (1990). At several places in this book we will pay attention to the subject of conditional error variance.

3.6 Estimating the true score

The true score can be estimated by the observed score, and so it is done frequently. Assuming that the measurement errors are approximately normally distributed, we can construct a 95% confidence interval:

$$x_p - 1.96\sigma_E \leq \hat{\tau}_p \leq x_p + 1.96\sigma_E \tag{3.9}$$

Unfortunately, the point estimate and the confidence interval in Equation 3.9 are misleading for two reasons. The first reason is that we can safely assume that the variance of measurement errors varies from person to person. Persons with a high or low true score have a relatively low error variance due to a ceiling and a floor effect, respectively. So, we should estimate error variance as a function of true score.

We will discuss the second reason in more detail. We start with a simple demonstration. Suppose all true scores are equal. Then the true-score variance equals zero. So, the observed-score variance equals the variance of measurement errors. We know this because we have obtained a reliability equal to zero. Which estimate of a person's true score seems most adequate? In this case, the best true-score estimate for all persons is the population mean μ_X.

More generally, we might estimate τ using an equation of the form $ax_p + b$, where a and b are chosen in such a way that the sum of the squared differences between true scores τ and their estimates are minimal. The resulting formula is the formula for the regression of true score on observed score:

$$\hat{\tau} = \frac{\sigma_T \rho_{XT}}{\sigma_X}(x - \mu_X) + \mu_T$$

This formula can be rewritten as follows:

$$\hat{\tau} = \rho_{XX'}x + (1 - \rho_{XX'})\mu_X \tag{3.10}$$

with a *standard error of estimation* (for estimating true score from observed score) equal to

$$\sigma_\varepsilon = \sigma_T \sqrt{1 - \rho_{XT}^2} = \sigma_X \sqrt{\rho_{XX'}} \sqrt{1 - \rho_{XX'}} = \sqrt{\rho_{XX'}} \, \sigma_E \qquad (3.11)$$

Formula 3.10 is known as the Kelley regression formula (Kelley, 1947). From Equation 3.11, it is clear that the Kelley estimate is better than the observed score as an estimate of true score.

The use of the Kelley formula can also be criticized:

1. The standard error of estimation (Equation 3.11) also supposes a constant error variance.
2. The true regression might be nonlinear.
3. The Kelley estimate of the true score depends on the population. Persons with the same observed score coming from different populations might have different true-score estimates and might consequently be treated differently.
4. The estimator is biased. The expected value of the Kelley formula equals τ_p only when the true score equals the population mean.
5. The regression formula is inaccurately estimated in small samples.

Under a few distributional assumptions, the Kelley formula can be derived from a Bayesian point of view. Assume that we have a prior distribution of true scores $N(\mu_T, \sigma_T^2)$—that is, the distribution is normal with mean μ_T and variance σ_T^2. Empirical Bayesians take the estimated population distribution of T as the prior distribution of true scores. Also assume that the distribution of observed score given true score τ equals $N(\tau, \sigma_E^2)$. Under these assumptions, the mean of the posterior distribution of τ given observed score x equals Kelley's estimate with μ_X replaced by μ_T. When a second measurement is taken, it is averaged with the first measurement in order to obtain a refined estimate of the true score. After a second measurement, the variance of measurement errors is not equal to σ_E^2 but is equal to $\sigma_E^2/2$. After k measurements, we have

$$\hat{\tau} = \frac{\sigma_T^2}{\sigma_T^2 + \sigma_E^2/k} x(k) + \frac{\sigma_E^2/k}{\sigma_T^2 + \sigma_E^2/k} \mu_T \qquad (3.12)$$

where $x(k)$ is the average score after k measurements, as the estimate of true score, and as k becomes larger, the expected value of Equation 3.12 gets closer to the value τ. So, the bias of the estimator does not seem to be a real issue.

3.7 Correction for attenuation

The correlation between two variables X and Y, ρ_{XY}, is small if the two true-score variables are weakly related. The correlation can also be small if one or both variables have a large measurement error. With the correlation being weakened or attenuated due to measurement errors, one might ask how large the correlation would be without errors (i.e., the correlation between the true-score variables). This is an old problem in test theory, and the answer is simple. The correlation between the true-score variables is

$$\rho_{T_X T_Y} = \frac{\sigma_{T_X T_Y}}{\sigma_{T_X} \sigma_{T_Y}} = \frac{\sigma_{XY}}{\sqrt{\rho_{XX'}} \sigma_X \sqrt{\rho_{YY'}} \sigma_Y} = \frac{\rho_{XY}}{\sqrt{\rho_{XX'}} \sqrt{\rho_{YY'}}} \qquad (3.13)$$

Formula 3.13 is the *correction for attenuation*. In practice, the problem is to obtain a good estimate of reliability. Frequently, only an underestimate of reliability is available. Then the corrected coefficient (Equation 3.13) can have a value larger than one in case the correlation between the true-score variables is high.

When data are available for several variables X, Y, Z, and so forth, we can model the relationship between the latent variables underlying the observed variables. In structural equation modeling, the fit of the structure that has been proposed can be investigated. So, structural equation modeling produces information on the true relationship between two variables.

Exercises

3.1 The reliability of a test is 0.75. The standard deviation of observed scores is 10.0. Compute the standard error of measurement.

3.2 The reliability of a test is 0.5. Compute test reliability if the test is lengthened with a factor $k = 2, 3, 4,..., 14$ ($k = 2(1)14$, for short).

3.3 Compute the ratio of the standard error of estimation and the standard error of measurement for $\rho_{xx'} = 0.5$ and $\rho_{xx'} = 0.9$. Compute the Kelley estimate of true score for an observed score equal to 30, and $\mu_X = 40$, $\rho_{xx'} = 0.5$, respectively, $\rho_{xx'} = 0.9$.

3.4 The reliability of test X equals 0.49. What is the maximum correlation that can be obtained between test X and a criterion? Explain your answer. Suggestion: Use the formula for the correction for attenuation.

3.5 Let ρ_{XY} be the validity of test X with respect to test Y. Write the validity of test X lengthened by a factor k, in terms of ρ_{XY}, σ_X, σ_Y, and $\rho_{XX'}$. What happens when k becomes very large?

CHAPTER 4

Estimating Reliability

4.1 Introduction

In this chapter, the major approaches to reliability estimation will be discussed. In Chapter 3, we noticed that test reliability is equal to the correlation between a test and a parallel test. The moment of administration of the second test, however, is of crucial importance, as it may have an influence on error variance. If there is a long time interval between the administration of the first and the second test, the factor of time may play an important role—persons may change in the time between testing. On the other hand, when tests are administered consecutively, fatigue is to be expected to come into play. Therefore, with test administration in one session, it is advisable to split the persons into two groups, one of which is administered test X first, followed by test Y, and the other is given the two tests in reverse order.

As the *parallel-test method* is not without its problems (see again Guttman, 1953), an alternative method for reliability estimation would be to administer the test twice. This method is the *test–retest method*. With a small time interval between test sessions, the risk is large that on the second test occasion persons remember their answers given on the first occasion. This would be a violation of the assumption of experimental independence. This violation would have a negative effect on the quality of the reliability estimate. With a larger time interval, persons might be changed on the characteristic of interest. Therefore, the test–retest method is useful only when a relatively stable characteristic is to be measured. The resulting reliability coefficient is called a stability coefficient for this reason.

There are also estimation methods based on data from a single administration of a test. These methods can be used when a test consists of several components, as most tests do. With these methods, the momentary level of achievement of a respondent is taken as the true score of interest. Consequently, a reliability coefficient obtained from the test at one occasion can differ from the stability coefficient.

Table 4.1 Major approaches to reliability estimation.

Reliability Coefficient	Major Error Source	Data-Gathering Procedure	Statistical Data Analysis
1. Stability coefficient (test–retest)	Changes over time	Test–retest	Product–moment correlation
2. Equivalence coefficient	Item sampling from test form to test form	Give form j and form k	Product–moment correlation
3. Internal consistency coefficient	Item sampling; test heterogeneity	A single administration	a) Split-half correlation and Spearman–Brown correction b) Coefficient alpha c) λ_2 d) Other

An overview of the major approaches to reliability estimation is given in Table 4.1.

In Section 4.2 estimation methods will be discussed based on a single administration of a test, in Sections 4.3 and 4.4 methods with parallel tests and test–retest approaches, in Section 4.5 reliability and factor analysis, in Section 4.6 the estimation of true scores and score profiles, and in Section 4.7 the conditional standard error of measurement.

4.2 Reliability estimation from a single administration of a test

When a test is composed of several parts, we might try to split the test into two parallel subtests. Then we might compute the correlation between the two halves. This correlation would give us an estimate of the reliability of a test with half the length of the original test. An estimate of the reliability of the original test can be obtained by applying the Spearman–Brown formula for a lengthened test. A weakness of the method is the arbitrary division of the test into two halves. This could easily be remedied by taking all possible splits into two halves. Should we confine ourselves to splits into two halves, however? The answer is no. Several coefficients have been proposed based on a split of a test into more than two parts (see Feldt and Brennan, 1989). We will discuss a method in which all parts or components play the same role.

Let test X be composed of k parts X_i. The observed score on the test can be written as

$$X = X_1 + X_2 + \cdots + X_k$$

and the true score as

$$T = T_1 + T_2 + \cdots + T_k$$

The reliability coefficient of the test is

$$\rho_{XX'} = \frac{\sigma_T^2}{\sigma_X^2} = \frac{\sum_i^k \sigma_{T_i}^2 + \sum_i^k \sum_{j \neq i}^k \sigma_{T_i T_j}}{\sigma_X^2}$$

The covariances between the true scores on the parts in the formula above equal the covariances between the observed scores on the parts. The true-score variances of the components are unknown. They can be approximated as described below.

While $(\sigma_{T_i} - \sigma_{T_j})^2 \geq 0$, we have

$$\sigma_{T_i}^2 + \sigma_{T_j}^2 \geq 2\sigma_{T_i}\sigma_{T_j}$$

We also have

$$\sigma_{T_i}\sigma_{T_j} \geq \sigma_{T_i T_j}$$

that is, the correlation coefficient does not exceed one, so

$$(k-1)\sum_{i=1}^k \sigma_{T_i}^2 = \sum_{i<j}^{k-1}\sum_j^k \left(\sigma_{T_i}^2 + \sigma_{T_j}^2 \right) \geq \sum_{i=1}^k \sum_{j \neq i}^k \sigma_{T_i T_j}$$

From this, we obtain

$$
\rho_{XX'} = \frac{\sum_{i=1}^{k}\sigma_{T_i}^2 + \sum_{i=1}^{k}\sum_{j\neq i}^{k}\sigma_{T_iT_j}}{\sigma_X^2} \geq \frac{\frac{k}{k-1}\sum_{i=1}^{k}\sum_{j\neq i}^{k}\sigma_{T_iT_j}}{\sigma_X^2}
$$

$$
= \frac{\frac{k}{k-1}\sum_{i=1}^{k}\sum_{j\neq i}^{k}\sigma_{X_iX_j}}{\sigma_X^2} = \left(\frac{k}{k-1}\right)\frac{\sigma_X^2 - \sum_{i=1}^{k}\sigma_{X_i}^2}{\sigma_X^2}
$$

Now we obtained a lower bound to the reliability, under the customary assumption of uncorrelated errors (for correlated errors, see Rae, 2006; Raykov, 2001). The coefficient is referred to as coefficient α:

$$
\alpha \equiv \left(\frac{k}{k-1}\right)\left(1 - \frac{\sum_{i=1}^{k}\sigma_{X_i}^2}{\sigma_X^2}\right) \tag{4.1}
$$

Coefficient α is also called a measure for internal consistency. We can elucidate the reason for this designation with an example. Let us take an anxiety questionnaire. Assume that different persons experience anxiety in different situations. Test reliability as estimated by coefficient α might be low, although anxiety might be a stable characteristic. The test–retest method might have given a much higher reliability estimate.

The popularity of coefficient α is due to Cronbach (1951). The coefficient was proposed earlier by Hoyt (1941) on the basis of an analysis of variance (see Chapter 5), and by Guttman (1945) as one of a series of lower bounds to reliability. Therefore, McDonald (1999, p. 95) refers to this coefficient as Guttman–Cronbach alpha. Following the *Standards* (APA et al., 1999), however, we will stick to calling it Cronbach's alpha. For dichotomous items, the item variance of item i can be simplified to $p_i(1 - p_i)$ if we divide by the number of persons N in the computation of the variances instead of $N - 1$. Here p_i is the proportion of correct responses to the item. The resulting coefficient is called Kuder–Richardson formula 20, KR20 for short (Kuder and Richardson, 1937). Kuder and Richardson proposed a further simplification, KR21. In KR21 all p_i are replaced by the average proportion correct. When the item difficulties are unequal, KR21 is lower than KR20. KR21 is discussed further in Chapter 6.

Under certain conditions, coefficient α is not a lower bound to reliability but an estimate of reliability. This is the case if all items (components) have the same true-score variance and if the true scores of the items correlate perfectly. In this case, the two inequalities in the derivation of the coefficient become equalities. Items or tests that satisfy this property are called (essentially) tau equivalent. The definition of *essentially tau-equivalent tests i* and *j* is

$$\tau_{ip} = \tau_{jp} + b_{ij} \tag{4.2}$$

If true scores are equal (i.e., if the additive constant b_{ij} equals 0), we have *tau-equivalent measurements*. Tau-equivalent tests with unequal error variances have unequal reliabilities. If true scores and error variances are equal, we have parallel tests. In the case of parallel test items, coefficient α can be rewritten in the form of the Spearman–Brown formula for the reliability of a lengthened test (Equation 3.7), where the reliability at the right-hand side of the equals sign (=) in the formula is replaced by the common intercorrelation between items.

A further relaxation of Equation 4.2 would be if the true scores of tests *i* and *j* are linearly related—that is, if

$$\tau_{ip} = a_{ij}\tau_{jp} + b_{ij} \tag{4.3}$$

In this case, we have the model of *congeneric tests*—true-score variances, error variances, as well as population means can be different. The congeneric test model is the furthest relaxation of the classical test model.

Let us have a further look at Equation 4.3. In Equation 4.3, the true score on test *i* is defined in terms of the true score on test *j*. An alternative and preferable formulation would be to write true scores on test *i* as well as test *j* in terms of a latent variable. So,

$$\tau_{ip} = a_i\tau_p + b_i \tag{4.4a}$$

and

$$\tau_{jp} = a_j\tau_p + b_j \tag{4.4b}$$

The true-score variances are $a_i^2\, \sigma_T^2$ and $a_j^2\, \sigma_T^2$. Without loss of generality, we can set σ_T^2 equal to one. For, if σ_T has a value u unequal to one, we can define a new latent score τ^* and new coefficients a^* with $\tau^* = \tau/u$ and $a^* = a \times u$, and the new latent score has a variance equal to one. The variances of the congeneric tests can be written as

$$\sigma_i^2 = a_i^2 + \sigma_{E_i}^2 \tag{4.5}$$

and the covariances as

$$\sigma_{ij} = a_i a_j \tag{4.6}$$

With three congeneric tests, there are three observed-score variances and three different covariances. There are six unknown parameters: three coefficients a and three error variances. The unknown parameters can be computed from the observed-score variances and covariances. With two congeneric tests, we have more unknowns than observed variances and covariances. In this case, we cannot estimate the coefficients a and the error variances. With more than three tests, more variances and covariances are available than unknown parameters. Then a statistical estimation procedure is needed in order to estimate the parameters from the data according to a specified criterion. Such a procedure is computer implemented in software for structural equation modeling (see Chapter 8).

It is important to have more than three tests when the congeneric test assumption is to be verified. (Three tests are enough to verify whether the stronger assumption of parallelism is satisfied.) The advantage of the exact computation of the coefficients a and the error variances in the case of three tests is apparent. Even when tests are not congeneric, it is possible to compute three values a for three tests, and in most cases, realistic error variances (with nonnegative values) are also obtained. With more than three tests, the assumption that tests are congeneric can be tested (Jöreskog, 1971). If the congeneric test model fits, we can also verify whether a more restrictive model—the (essentially) tau-equivalent test model or the model with parallel tests—fits the data. If a simpler, more restrictive model adequately fits the data, this model is to be preferred. It is also possible that the congeneric model does not fit. Then we can try to fit a structural model with more than one dimension (Jöreskog and Sörbom, 1993).

The administration of a number of congeneric tests is practically unfeasible. However, an existing test might be composed of subtests that are congeneric, tau-equivalent, or even parallel. In such a situation, the method for estimating coefficients a for congeneric measurements can be used for the estimation of test reliability. If we have congeneric subtests, the estimate of reliability is

$$\rho_{XX'} = \frac{\left(\sum_{i=1}^{k} a_i \right)^2}{\sigma_X^2} \tag{4.7}$$

If coefficients a and error variances of the subtests are available, it is possible to use them for computing weights that maximize reliability. Jöreskog (1971; see also Overall, 1965) demonstrated that with congeneric measurements, optimal weights are proportional to

$$w_i = \frac{a_i}{\sigma_{E_i}^2} \tag{4.8}$$

In other words, the optimal weight is smaller for a large error variance and higher in case the subtest contributes more to the true score of the total test. More information on weighting is given in Exhibit 4.1.

Exhibit 4.1 Weighting responses and variables

A total score is obtained by adding item scores. The total score can be an unweighted sum of the item scores or a weighted sum score. Two kinds of weights are in use: a priori weights and empirical weights. Empirical weights are somehow based on data. Many proposals for weighting have been done. Among these proposals are the optimal weights for congeneric measurements and weights that are defined within the context of item response theory.

We mention one other proposal for weights here—the weighting of item categories and items on the basis of a homogeneity analysis. Homogeneity analysis is used for scaling variables that are defined on an ordinal scale. Weights are assigned to the categories of these variables. The weights and scores have symmetrical roles. A person's score is defined as the average of the category weights of the categories that were endorsed. The category weight of a variable is proportional to the average score of the persons who chose the category. Actually, one of the algorithms to obtain weights and scores is to iterate between computing scores on the basis of the weights and weights on the basis of the scores until convergence has been reached.

Lord (1958) has demonstrated that homogeneity analysis weights maximize coefficient alpha. In the sociological literature, coefficient alpha with optimally weighted items is known as theta reliability (Armor, 1974). Further information on alpha and homogeneity analysis can be found in Nishisato (1980). A more readable introduction into homogeneity analysis (or dual scaling, optimal scaling, correspondence analysis) is provided by Nishisato (1994).

The so-called maxalpha weights are optimal weights within the context of homogeneity analysis. In other approaches other weights are found to be optimal. A general treatment of weighting is given by McDonald (1968). When items are congeneric, the weights that maximize reliability are obviously optimal, and these weights are not identical to the maxalpha weights. The ultimate practical question is this: Is differential weighting of responses and variables worth the trouble? In the context of classical test theory, the answer is "seldom." Usually, items are selected that are highly correlated. Then the practical significance is limited (cf. Gifi, 1990, p. 84).

Let us now return to coefficient α and the question of alpha as a lower bound to the reliability of a test. For a test composed of a reasonably large number of items that are not too heterogeneous, coefficient α slightly underestimates reliability. On the other hand, it is possible for coefficient α to have a negative value, although reliability—being defined as the ratio of two variances—cannot be negative. Better lower bounds than coefficient α are available. Guttman (1945) derived several lower bounds. One of these, called λ_2, is always equal to or larger than coefficient α. The formula for this coefficient is

$$\lambda_2 \equiv \frac{\sigma_X^2 - \sum_{i=1}^{k} \sigma_{X_i}^2 + k\sqrt{\dfrac{1}{k(k-1)}\sum_{i=1}^{k}\sum_{j\neq i}^{k} \sigma_{X_i X_j}^2}}{\sigma_X^2} \qquad (4.9)$$

An example of reliability estimation with a number of coefficients is presented in Exhibit 4.2.

Exhibit 4.2 An example with several reliability estimates

Lord and Novick (1968, p. 91) present the variance–covariance matrix for four components, based on data for the Test of English as a Foreign Language. Their data are replicated in the table below. From the table, we can read that the variance of the first component equals 94.7; the covariance between components 1 and 2 equals 87.3.

	C_1	C_2	C_3	C_4
C_1	94.7	87.3	63.9	58.4
C_2	87.3	212.0	138.7	128.2
C_3	63.9	138.7	160.5	109.8
C_4	58.4	128.2	109.8	115.8

We use the data in the table for the computation of several reliability coefficients. First, let us compute split-half coefficients with a Spearman–Brown correction for test length. The total test can be split into two half tests in three different ways. We compute all three possible reliability estimates.

Split(a,b)	Var(a)	Var(b)	Cov(a,b)	r	$r(2)$
12–34	481.30	495.90	389.20	0.797	0.887
13–24	383.00	584.20	394.20	0.833	0.909
14–23	327.30	649.90	389.20	0.844	0.915

The estimates vary from 0.887 to 0.915. An alternative approach on the basis of the split of the test into two halves would have been to use coefficient alpha with two components.

Next we compute coefficient alpha. The total score variance is equal to the sum of all cell values in the table: 1755.6. The sum of the component variances equals 583.0. Coefficient alpha equals $\alpha = (4/3)(1 - 583.0/1755.6) = 0.891$.

The value of α is lower than the highest estimate based on a split into two parts. Coefficient alpha is guaranteed a lower bound to reliability, the split-half coefficient is not. The most adequate estimate based on splitting the test into halves seems to be the first, because the split 12–34 seems to produce more or less comparable halves.

Finally, we compute λ_2. We need the square root of the average value of the squared covariances: 102.3426. We obtain $\lambda_2 = (1755.6 - 583.0 + 4 \times 102.3426)/1755.6 = 0.901$.

The value of λ_2 is higher than the value of α.

It is worthwhile to discuss two other lower bounds. The first is the g.l.b., the "greatest lower bound"; its definition will be discussed in Section 4.5. The second is coefficient α_s, the stratified coefficient α (Rajaratnam, Cronbach, and Gleser, 1965).

First, let us rewrite coefficient α as

$$\alpha = \frac{k^2 \text{ave}(\sigma_{ij})}{\sigma_X^2} \tag{4.10}$$

where ave denotes average and σ_{ij} is shorthand for the covariance between item i and item j. Figure 4.1 illustrates the situation for a four-item test. The diagonal entries in the figure represent the item variances. The off-diagonal entries represent the covariances between items. The sum of the entries equals the variance of the total test, the denominator in Equation 4.10. The numerator of coefficient α according to Equation 4.10 is obtained by replacing all diagonal values in the figure by the average covariance and, next, summing all entries.

σ_1^2	σ_{12}	σ_{13}	σ_{14}
σ_{21}	σ_2^2	σ_{23}	σ_{24}
σ_{31}	σ_{32}	σ_3^2	σ_{34}
σ_{41}	σ_{42}	σ_{43}	σ_4^2

Figure 4.1 The variance–covariance matrix for a four-item test.

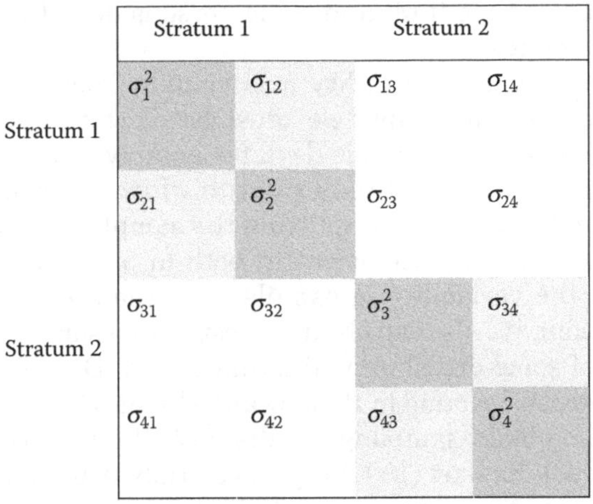

	Stratum 1		Stratum 2	
Stratum 1	σ_1^2	σ_{12}	σ_{13}	σ_{14}
	σ_{21}	σ_2^2	σ_{23}	σ_{24}
Stratum 2	σ_{31}	σ_{32}	σ_3^2	σ_{34}
	σ_{41}	σ_{42}	σ_{43}	σ_4^2

Figure 4.2 The variance–covariance matrix of a four-item test with two strata.

Now suppose that we can classify the items into two relatively homogeneous clusters or strata. We can use this stratification in the computation of the estimated total true-score variance. We can replace the item variances within a stratum by the average covariance between items belonging to this stratum instead of by the average covariance computed over all item pairs. So, in the example in Figure 4.2, the variances of items 1 and 2 are replaced by σ_{12} (= σ_{21}).

The stratified coefficient alpha can be written as

$$\alpha_s = \frac{\sum_{i=1}^{q} \alpha(i)\sigma_{Y_i}^2 + \sum_{i=1}^{q}\sum_{j \neq i}^{q} \sigma_{Y_iY_j}}{\sigma_X^2} \qquad (4.11)$$

where q is the number of strata, Y_i the observed score in stratum i, and $\alpha(i)$ coefficient α computed over the items in stratum i. A more general reliability formula from test theory is obtained if we replace $\alpha(i)$ in Equation 4.11 by a possibly different reliability estimate for subtest i.

Reliability estimation based on a measure of internal consistency is problematic in case the item responses cannot be considered experimentally independent. This might happen, for example, if the test is answered under a time limit and some persons do not reach the items at the end of the test.

We always estimate reliability in a sample from the population of interest. With a small sample we must be alert to the risk that the reliability estimate in the sample deviates notably from the value in the population. An impression of the extent to which the sample estimates might vary can be obtained by splitting the sample into two halves and computing the reliability coefficient in both (a procedure that gives an impression of the variability in samples half the size of the sample in the investigation). We also can obtain an estimated sampling distribution on the basis of some distributional assumptions. Distributional results for coefficient α can be found in Pandey and Hubert (1975), among others. One might also obtain sampling results with the bootstrap (Efron and Tibshirani, 1993). Raykov (1998) reports a study using the bootstrap for obtaining the standard error for a reliability coefficient.

4.3   Reliability estimation with parallel tests

Interchangeable test forms in terms of the definition of parallelism are used for the parallel-forms or alternate-forms method of assessing reliability. It has been noticed that parallel tests are not uniquely defined. The same test could belong to more than one set of parallel tests, leading, in general, to more than one reliability coefficient.

A practical problem of the parallel-forms approach is that the instruments might not satisfy the requirements of parallel tests. Then alternative equivalence models as given in Section 4.2 could be considered.

In estimating reliability for, say, test forms X and Y, special attention must be paid to the design of the reliability study. We must take refuge in balanced designs to eradicate as best as we can possible order effects. In addition, we may consider what the time interval should be between the presentation of the first and the second test forms.

4.4   Reliability estimation with the test–retest method

This method for assessing reliability is straightforward enough: present the test at two occasions to the same group of subjects and correlate the outcomes.

Assessing reliability with the test–retest method has its problems. What, for example, occurred between occasions 1 and 2? Are subgroups perhaps affected differentially by influences (e.g., in educational testing where the test is administered in several classes and classes are treated differently)? Also, memory effects may influence reliability estimates.

Table 4.1 lists the reliability coefficient of the test–retest method as a stability coefficient, but what about the test–retest approach for the measurement of change? To be sure, the measurement of change is a legitimate research subject of its own.

Retesting would perhaps be the most appropriate method for the estimation of the reliability of speeded tests. With speeded tests, internal consistency estimates of reliability are inappropriate.

4.5 Reliability and factor analysis

The analysis with congeneric measurements in Section 4.2 is an example of a linear factor analysis—to be precise, a factor analysis with one common factor. While there are no replications, the possible contribution of unique factors to observed scores is subsumed under the error component. Generally, one hopes for one dominant common factor, indicating that one is measuring a single construct. Frequently, however, a factor analysis results in more than one common factor. Then the true score can be written as a weighted sum of factor scores (see Chapter 8, especially Figure 8.4).

In a factor analysis, several choices have to be made that influence the sizes of the estimated error variances of the variables and, consequently, the size of the reliability coefficient that can be computed. After all, reliability equals one minus the sum of the error variances divided by the observed-score variance.

The greatest lower bound (g.l.b.) to reliability can be obtained from a special kind of factor analysis, constrained minimum trace factor analysis, in which the sum of the error variances is as large as possible given the variances of the subtests and their covariances. The determination of the g.l.b. is a complicated problem (Ten Berge and Sŏcan, 2004).

4.6 Score profiles and estimation of true scores

Sometimes a test is heterogeneous, and on the basis of a factor analysis several subtests can be discerned. If subtests are defined, we can compute a total score as well as a separate score for each of the subtests.

The higher the correlation between subtests and the less reliable sub-tests are, the less useful it is to compute subtest scores in addition to a total score. With reliable subtests that do not correlate too strongly, it makes sense to compute subtest scores in addition to or instead of total test score.

With subtest scores, we can compute a score profile for each person tested. We can verify whether a person has relatively strong and relatively weak points. We can determine to what extent the score profile of a person is deviant. For a solid interpretation of the profile of scores, it is important to standardize the subtests so that they have the same score distribution in the relevant population of persons. The subtests should have identical means and standard deviations (for norms with respect to the estimation of means see Angoff, 1971; for sampling techniques, see Kish, 1987). Only then is it relatively simple to notice whether a person scores relatively high on one subtest and relatively low on another (e.g., relatively high on a verbal subtest and relatively low on a mathematical subtest). When subtest reliabilities vary notably, the advantage of this way of scaling the subtests is limited however, for then there are large differences between the true-score scales (Cronbach et al., 1972).

The observed score on a subtest is an obvious estimate of the true score on the particular subtest. In a previous section it was demonstrated that the observed score was not optimal: the Kelley estimate performs better than the observed score. For profile scores one can think of generalizing Kelley's formula.

Let us take a profile with two subtests X and Y as an example. We are interested in the true score of a person p on subtest X, $\tau_p(X)$. If we knew the true scores on test X, we would certainly consider the possibility to "predict" these scores from observed scores x and y using multiple regression. There is no reason not to use the multiple regression formula in case the criterion τ is unknown. The formula with which we "predict" the true score on subtest X is

$$\hat{\tau}_p(X) = \mu_{T(X)} + \frac{\sigma_{T(X)}}{\sigma_X} \frac{\rho_{T(X)X} - \rho_{T(X)Y}\rho_{XY}}{1 - \rho_{XY}^2}(x_p - \mu_X)$$

$$+ \frac{\sigma_{T(X)}}{\sigma_Y} \frac{\rho_{T(X)Y} - \rho_{T(X)X}\rho_{XY}}{1 - \rho_{XY}^2}(y_p - \mu_Y)$$

(4.12)

In this formula several correlations with true scores on X are involved, and these correlations are unknown. Also unknown is the standard deviation of true scores on subtest X. However, all the unknowns can be estimated:

$$\sigma_{T(X)} = \sqrt{\rho_{XX'}}\,\sigma_X$$

$$\rho_{T(X)X} = \sqrt{\rho_{XX'}}$$

and, analogous to the correction for attenuation for two variables X and Y,

$$\rho_{T(X)Y} = \frac{\rho_{XY}}{\sqrt{\rho_{XX'}}}$$

We can conclude that the best estimate (in the least squares sense) of the true score on subtest X makes use of the score on subtest Y as well. With reliable X, the score on subtest X gets a high weight. The weight of test X is also relatively high in case the scores on subtest Y are nearly uncorrelated with those on subtest X. In case true scores (and observed scores) on test Y are uncorrelated with those on X, the formula can be simplified to the Kelley formula. With congeneric subtests X and Y, the obtained weights equal the optimal weights for congeneric measurements (Equation 4.8). In case true scores on subtests X and Y are strongly correlated and subtest X is relatively unreliable, it is possible to have a smaller weight for X than for Y in the formula for the prediction of the true score on X.

It is instructive to write Equation 4.12 in terms of variances and covariances:

$$\hat{\tau}_p(X) = \mu_X + \frac{\sigma_{T(X)X}\sigma_Y^2 - \sigma_{T(X)Y}\sigma_{XY}}{\sigma_X^2\sigma_Y^2 - \sigma_{XY}^2}(x_p - \mu_X)$$

$$+ \frac{\sigma_{T(X)Y}\sigma_X^2 - \sigma_{T(X)X}\sigma_{XY}}{\sigma_X^2\sigma_Y^2 - \sigma_{XY}^2}(y_p - \mu_Y)$$

(4.13)

We can rewrite this equation as follows:

$$\hat{\tau}_p(X) = \hat{\tau}_p(X)\,|\,y_p + \frac{\sigma_{T(X)}^2 - \sigma_{T(X)Y}^2 / \sigma_Y^2}{\sigma_X^2 - \sigma_{T(X)Y}^2 / \sigma_Y^2}(x_p - \hat{\tau}_p(X)\,|\,y_p)$$ (4.14)

where

$$\hat{\tau}_p(X) \mid y_p = \mu_X + \frac{\sigma_{T(X)Y}}{\sigma_Y^2}(y_p - \mu_Y) \qquad (4.15)$$

In other words, the optimal prediction formula for predicting the true score on X given observed scores on X and Y can be viewed as a two-step process. First, we estimate the true score on X given the observed score on Y. Next, we improve this estimate using the information given by the observed score on X. This way of viewing the estimation procedure would be quite logical if we take different measurements at different occasions. For example, Y might be the first measurement with a measurement instrument and X the second. In the Kalman filter, the estimate of true score on time t is based on test data obtained at time t and the true-score estimate at time $t-1$ (Oud, Van den Bercken, and Essers, 1990).

The estimation of profile scores with Equation 4.12 can evoke similar objections as the application of Kelley's formula in connection with a single test. The estimate of a person's true score depends on the population that serves as a reference. Certainly, when we use profile scores, it is obvious that we compare the outcomes for a person with results for a reference population whether we use Kelley's formula or not. The subtests are scaled in such a way that they have the same mean in some population. And when more than one relevant population exists, there is nothing against making separate norms for these different populations.

Another disadvantage of the use of a formula like Equation 4.12 seems to be the detection of persons with deviant score patterns. Let two subtests correlate strongly. The estimation formula then gives similar estimates of the two true scores. The relevant information that the pattern of scores is deviant is likely to be missed.

We can find out whether a score pattern is aberrant. We will demonstrate this with observed scores on two tests X and Y. The prediction of the score on test Y, given the score on test X, is given by the regression equation:

$$\hat{Y} = \sigma_Y \rho_{XY}(X - \mu_X)/\sigma_X + \mu_Y \qquad (4.16)$$

with a standard error of prediction equal to

$$\sigma_\varepsilon = \sigma_Y \sqrt{1 - \rho_{XY}^2} \qquad (4.17)$$

We can compute the predicted value on test Y and construct a 95% confidence interval using the assumption of normally distributed prediction errors. If the observed score on test Y lies outside this interval, we have an argument to consider the score pattern as aberrant. We might also evaluate the raw score difference $x - y$. Then we evaluate the difference irrespective of the correlation between X and Y. The relevant standard deviation for the raw-score difference is

$$\sigma_{E(X-Y)} = \sqrt{\sigma_{E(X)}^2 + \sigma_{E(Y)}^2} = \sqrt{\sigma_X^2(1-\rho_{XX'}) + \sigma_Y^2(1-\rho_{YY'})} \qquad (4.18)$$

With Equation 4.18 we can construct a 95% confidence interval for the difference between true scores on tests X and Y. When the true scores on tests X and Y have a correlation smaller than one, then more, perhaps much more than 5%, of the observed differences fall outside of this interval.

A special application of profiles is that in which scores X and Y are two measurements on the same measurement instrument, taken at two different occasions. Now we might be interested in the possibility of a true-score change or a true-score gain. The simplest way to estimate the true difference is to use the difference score. However, difference scores have a bad reputation. They can be quite unreliable even in case the separate measurements are highly reliable. Difference scores are used when the two measurements are related. So, we may assume that the true scores on both measurements are strongly correlated. Suppose that we have a situation in which the true scores on both measurements are equal. Then the true change is zero, and the reliability of difference scores is zero, too.

On the other hand, a low reliability does not imply that there are no changes. It is possible that all persons have changed the same amount between testing occasions. On the group level, the measurement of change is useful even with a low reliability for difference scores.

The presence of measurement error affects change in a special way. Let us analyze this in a simple situation in which the mean and variance of scores are equal in a pretest and a posttest. We will notice that there are changes although there is no overall change. The persons with better-than-average scores on the pretest will on the average have lower scores on the posttest. Persons with lower-than-average scores on the pretest will on the average show some score gain. The scores regress to the mean. This effect appears even if there is no true change. The effect is due to measurement error. Among the high scores on the pretest, there are always some scores that are high due to measurement error. Among the low scores on the pretest, there are always low scores due to measurement error. The difference score (posttest–pretest) is negatively correlated with the measurement error on the pretest. The true difference is better estimated by equations like Equation 4.12 (Lord and Novick, 1968, pp. 74–76).

Due to the regression to the mean, the use of difference scores in research is problematic. For research, alternatives are available (see Burr and Nesselroade, 1990; Cronbach and Furby, 1970). Rogosa, Brandt, and Zimowski (1982) discuss the possibility of modeling growth in case of more than two occasions.

4.7 Reliability and conditional errors of measurement

The 1985 as well as the 1999 *Standards* emphasized to report reliability as well as the standard errors of measurement. And, in addition, Standard 2.2 of the 1999 edition states:

> The standard error of measurement, both overall and conditional (if relevant), should be reported both in raw scores or original scale units and in units of each derived score recommended for use in test interpretations. (*Standards*, APA et al., 1999, p. 31)

The standard error of measurement can vary with true-score level. Conditional standard errors of measurement are standard errors of measurement conditional on true-score level. Such standard errors of measurement can be used as an alternative approach to convey reliability information, by constructing a confidence interval for an examinee's true score, universe score (to be discussed in Chapter 5), or percentile rank. Earlier, three types of standard errors were discussed: the standard error of measurement (Equation 3.3), the standard error of estimate of true score (Equation 3.11), and the standard error of prediction (Equation 4.17).

Woodruff (1990) studied the conditional standard error of measurement for assessing the precision of a test on its score scale. He proposed to split a test into two parallel halves X and X'. ANOVA is used to estimate values $\sigma^2(E'|X)$ as substitutes of $\sigma^2(E'|T)$. Then the outcomes are corrected for the fact that the test was split into two halves (using the customary assumption that the error variance doubles for a test lengthened by a factor 2).

Feldt and Qualls (1996) proposed a method for the estimation of the conditional error variance based on a split of the test into a number of essentially tau-equivalent subtests. It is possible to use a split of the test into two halves, but it proves to be better to split the test in many subtests as long as all subtests can be considered as essentially tau-equivalent measurement instruments. Let there be n subtests. For person p, the estimated error variance of the subtests is

$$s^2_{E(p)} = \frac{\sum_{i=1}^{n}[(x_{pi}-x_{p.})-(x_{.i}-x_{..})]^2}{n-1} \tag{4.19}$$

where the scores are corrected for the test effects $x_{.i} - x_{..}$. In the terminology of ANOVA, two-way interactions are used in Equation 4.19. Suppose that the subtests have equal score ranges. Then the consequence of the assumption of essentially tau-equivalent subtests on which Equation 4.19 is based, is that the error variance associated with a perfect score is nonzero when the subtests differ in difficulty level. In a nonlinear true-score model, a model based on item response theory (IRT), such a strange effect does not occur.

Again, we must multiply the estimate with a constant in order to obtain the error variance on the total test. When the n subtests add up to the total test, total test length is n times the length of the subtests and the result in Equation 4.19 must be multiplied by n.

Next, the error variances for all persons with the same total score can be averaged. This produces the estimated relationship between the size of the conditional error variance and total score. Feldt and Qualls suggest to reduce sampling variation further by smoothing the empirical relationship between error variance and total score. This can be achieved by a polynomial regression, where the error variance is regressed on powers of X (X, X^2, etc.).

It might be interesting to compare Equation 4.19 with a formula for the conditional error variance developed within the context of generalizability theory. For this purpose, Equation 4.19 is rewritten as

$$s^2_{E(p)} = s^2(x_{pi} \mid p) + s^2(x_{.i}) - 2\text{cov}(x_{pi}, x_{.i} \mid p) \tag{4.20}$$

which is comparable to Equation 5.41.

More methods for estimating conditional standard errors of measurement are described by Lee, Brennan, and Kolen (2000). Methods for obtaining conditional error variances have been proposed specifically within the context of generalizability theory (Chapter 5) and for dichotomous items (Chapter 6). In IRT, the problem of the conditional standard error of measurement is approached in another way (see Section 6.4).

Exercises

4.1 A test X is given with three subtests, X_1, X_2, and X_3. The variance–covariance matrix for the subtests is given in the table below. Estimate reliability with coefficient α.

	X_1	X_2	X_3
X_1	8.0	6.0	8.0
X_2	6.0	12.0	12.0
X_3	8.0	12.0	17.0

4.2 Use the variance–covariance matrix from Exercise 4.1 for estimating test reliability according to the model of congeneric tests. Use Equation 4.6 for the estimation of the a_i.

4.3 Prove that for parallel test items coefficient alpha equals the Spearman–Brown formula for the reliability of a lengthened test.

4.4 Two tests X_1 and X_2 are congeneric measurement instruments. Their correlations with other variables Y_1, Y_2, and so on, differ. Is there a pattern to be found in the correlations?

4.5 Given are two tests X and Y with $\sigma_X^2 = 16.0$, $\sigma_Y^2 = 16.0$, $\rho_{XX'} = \rho_{YY'} = 0.8$, and $\rho_{XY} = 0.7$.
   a. Compute the observed-score variance, the true-score variance, and the reliability of the difference scores $X - Y$.
   b. Compare the variance of the raw score differences with $\sigma_{E(X-Y)}^2$ of Equation 4.18.

4.6 In a test, several items cover the same subject. Which assumption of classical test theory might be violated? What should we do when we want to estimate reliability with coefficient α?

4.7 We have three tests X_1, X_2, and X_3 measuring the same construct. Their correlations with test Y equal 0.80, 0.70, and 0.60. Their covariances with Y are equal to 0.20. The means of the tests are 16.0, 16.0, and 20.0, respectively. Are these tests parallel tests, tau-equivalent, essential tau-equivalent, or congeneric? Discuss your answer.

4.8 A test has a mean score equal to 40.0, a standard deviation equal to 10.0, and a reliability equal to 0.5. Which difference score do you expect after a retest when the first score of a person equals 30?

4.9 Two tests X and Y are available. The tests have equal observed-score variances: $\sigma_X^2 = \sigma_Y^2 = 25.0$. The reliability of test X is 0.8, the reliability of test Y is 0.6. Their intercorrelation is zero. Compute the reliability of the composite test $X + Y$. Also, compute the reliability of the composite after doubling the test length.

CHAPTER 5

Generalizability Theory

5.1 Introduction

An observed score obtained with a measurement instrument is just
one of many possible scores that could have been obtained, because
an alternative context, other measurement conditions, and varying
circumstances may lead to other observed scores. In other words, an
observed score is usually obtained for a particular test form. Another
equivalent test form, however, may have been as appropriate for our
measurement purpose but might have led to a different observed test
score. Consequently, if one wants to model observed scores, one has to
take into account many sources of variation (including error variation).
This also applies if one considers the reliability of scores obtained from
a measurement instrument. Classical reliability provides a decomposition of the observed score into a true score and only one type of error.
Theoretically, this error is undifferentiated. Several reliability estimation procedures lead to specific conceptualizations of error: parallel
test forms reliability e.g., considers the lack of equivalence between
the forms as the source of error, test–retest reliability the time of
testing, and internal consistency reliability the variability in test
items.

Generalizability theory or G theory (Cronbach, Rajaratnam, and
Gleser, 1963; Cronbach et al., 1972; Brennan, 2001), in contrast to classical test theory, provides a decomposition of an observed score taking
into account more sources of variation, dependent upon the specific measurement situation. G theory also recognizes that multiple sources of
error may operate in a measurement, implying that there is no unitary
definition of reliability. This is, basically, a consequence of the view that
a specific test score or any other particular behavioral measurement, for
that matter (e.g., a job performance score or an expert's performance
assessment of student achievement), is conceived of as a sample from a
universe of admissible or suitable observations. Such a universe is characterized by one or more sources of variation, the *facets*. In Section 5.2

an overview of the basic concepts of G theory will be given. In view of how a particular behavioral measurement is conceived, the designs used to collect the measurements come in focus. In Sections 5.3 and 5.4, some of the more simple one- and two-facet designs will be given, together with the corresponding decomposition of the observed score into a component for the true score, viz. the universe score, and one or more error components. We will see that with each of the components, variances, or rather variance components, are associated. In Section 5.5 an extensive example will be described from a study by Webb, Shavelson, Kim, and Chen (1989) on the reliability of job performance measurements. In Sections 5.6 and 5.7, two-facet nested designs are introduced. In Section 5.8 designs with fixed facets are discussed, and in Section 5.9 kinds of measurement errors. In Section 5.10 attention is paid to conditional errors of measurement. Finally, in Section 5.11 some concluding remarks are made.

5.2   Basic concepts of G theory

A particular behavioral measurement is conceived of as a sample from a *universe* of admissible observations, or a domain of suitable or appropriate observations. The universe or domain is characterized by one or more sources of error variation, called *facets*. In a study where students' performance is rated by judges on performance criteria, judges and performance criteria are the facets of a measurement, and each facet consists of a set of conditions. Usually the facets are assumed to be indefinitely large. The universe, then, is defined as all possible conditions of the facets. Ideally, a person's *universe score* is his or her average score over all conditions. In a measurement situation, however, error is at stake, and this calls for the estimation of variance components. These estimates of variance components are highly informative, whereas the so-called *generalizability coefficients*—the G theory counterparts of classical reliability coefficients—are straightforward ratios of appropriate universe-score variances to total-score variances. In addition to variance components, standard errors are considered as appropriate indicators of uncertainty of, for example, performance assessments of student achievement or school effectiveness (Cronbach, Linn, Brennan, Haertel, 1997).

The purpose of G theory is to generalize from an observation at hand (i.e., the observed score) to the appropriate universe of observations. This domain or universe is defined by all possible conditions of the facets of the study. It should be noticed that the *object of measurement* (e.g., the persons to whom a test is presented) is not a facet.

Judge 1		Judge 2	
Item 1	Item 2	Item 1	Item 2
Person 1 ×	×	×	×
Person 2 ×	×	×	×

$p \times i \times j$

Judge 1		Judge 2	
Item 1	Item 2	Item 3	Item 4
Person 1 ×	×	×	×
Person 2 ×	×	×	×

$p \times (i:j)$

Figure 5.1 A crossed $p \times i \times j$ design and a nested $p \times (i:j)$ design with the same number of measurements for each person.

The measurements in a study are obtained according to a particular design. A one-facet example of a measurement design is an n-item test presented to a number of persons. In this example, we have a crossed design with all combinations of items (i) and persons (p); this crossed design is denoted as $p \times i$. If different sets of items are presented to different persons, we have a nested $i:p$ (read i within p) design instead. With two facets, several (partially) nested designs are possible. Let us take a study with items (i) and raters or judges (j). In this study all persons (p) have answered all items. This part of the study is a crossed $p \times i$ design. The items have been distributed among the judges. In other words, each judge had another set of items to rate: Items are nested within judges. Judges have been crossed with persons: If an item has been allocated to a judge, he or she has rated the answers of all persons to this item. This partially nested design is denoted as $p \times (i:j)$. A representation of a nested $p \times (i:j)$ design and that of a crossed $p \times i \times j$ design with the same number of measurements for each person is given in Figure 5.1.

In G theory two types of facets are distinguished: *random* facets and *fixed* facets. In a random facet the number of conditions is thought to be infinite. That is, the conditions of a facet that are selected for a particular measurement procedure or test are assumed to be a random sample from a very large number of possible conditions. And we would like to generalize over all admissible or suitable observations from the universe. Over fixed facets no generalization is sought, as the number of conditions in a fixed facet is equal to the number of conditions in the study or measurement procedure. A combination of a random facet and a fixed facet leads to a mixed-facet generalizability study.

In G theory a distinction is made between a generalizability or G study, and a decision or D study. A G study investigates the influence of the sampling of conditions from various facets on observed scores.

So the purpose of a G study is to collect as much information on the sources of variation as possible. A D study provides information for substantive decisions. It decides on the specific design that is most suitable for typical applications. Clearly, such a decision depends upon several factors. First, the relevant universe of generalization must be defined. Second, it must be decided whether the purpose of the study involves relative or absolute decisions. Relative decisions refer to the comparison of a person's achievement (e.g., relative to other persons' achievements). Absolute decisions are made when one is interested in a person's universe score, per se. With relative decisions go relative errors, and absolute errors are associated with absolute decisions. A third factor upon which the choice of a D study depends is the size of the various sources of variation. In addition, practical considerations such as the availability of judges (e.g., in a rating study) and costs associated with gathering the data are criteria for deciding on a specific measurement design. As in a D study, alternative possibilities can be tried out with respect to the number of conditions of the facets involved. A D study can be regarded as a generalization of the Spearman–Brown formula for test length.

5.3   One-facet designs, the $p \times i$ design and the $i : p$ design

5.3.1   The crossed design

Let us have a universe with one facet, the facet *items*. We assume that the facet is a random facet. For person p, the observed score on item i is X_{pi}. When we want to generalize over the facet, we must take the expectation of X_{pi} over items. This expectation defines the universe score:

$$\mu_p \equiv \mathrm{E}_i X_{pi} \qquad (5.1)$$

The universe score is comparable to the true score in classical test theory. In generalizability theory, it is assumed that the persons are a random sample from a large population (formally: $N_p = \infty$). Analogous to Equation 5.1, we can define the population mean of item i as

$$\mu_i \equiv \mathrm{E}_p X_{pi} \qquad (5.2)$$

The expectation of the universe scores is μ. This universe mean is also the expectation of the population means μ_i.

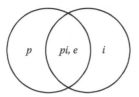

Figure 5.2 Venn diagram representation of the $p \times i$ design.

With these definitions, we can decompose the observed score X_{pi} into a number of components:

$$
\begin{array}{lll}
X_{pi=} & \mu & \text{Grand mean} \\
& + \mu_p - \mu & \text{Person effect} \quad (5.3) \\
& + \mu_i - \mu & \text{Item effect} \\
& + X_{pi} - \mu_p - \mu_i + \mu & \text{Residual}
\end{array}
$$

So, the observed score can be written as the sum of the grand mean, a person effect, an item effect, and a residual. The residual can be thought of as a combination of pure measurement error and the interaction between the item and the person. But, for lack of replications, these two sources of variation are confounded.

In Figure 5.2 a Venn diagram representation is given of the $p \times i$ design. An advantage of such a representation is that it visualizes the variance components involved in the decomposition of the observed scores. The interaction can be found in the segment where the circles for persons and items overlap.

The population variance of universe scores or person effects is called the variance component for persons and is written as σ_p^2. We also have a variance component for items, σ_i^2, and a residual variance component, $\sigma_{pi,e}^2$. The notation of the residual reflects the confounding of the random error and the interaction. The variance of X_{pi} over p and i, $E_{p,i}(X_{pi} - \mu)^2$ is

$$
\sigma^2(X_{pi}) = \sigma_p^2 + \sigma_i^2 + \sigma_{pi,e}^2 \tag{5.4}
$$

The three variance components can be estimated from an analysis of variance (ANOVA) of a two-way design. Statistical packages are available for ANOVA analyses and the estimation of variance components

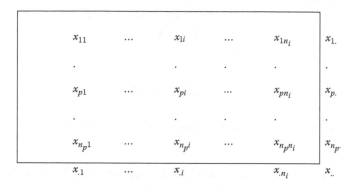

Figure 5.3 Observations in a crossed $p \times i$ design with n_p persons and n_i items.

(e.g., SAS, www.sas.com; SPSS, www.spss.com). There also is a suit of computer programs for generalizability analyses, Genova, which can be obtained for free from the University of Iowa, College of Education (Brennan, 2001).

It should be noticed that in the ANOVA terminology, two ways are distinguished: one represents the units of measurements (i.e., the persons in G theory), and the other the facet, items. So, a two-way ANOVA design is equivalent to or rather leads to a one-facet G study. The observations in a crossed design can be written as in Figure 5.3. In the rightmost column, we have the averages for persons, averaging over items. In the bottom row, we have the average scores for items, averaging over persons.

In order to compute the variance components for this $p \times i$ design, we use the ANOVA machinery. We start with calculating the sums of squares for persons and items, respectively. For the computation of the sum of squares for persons, we replace each entry x_{pi} in the row for a person by the average score for this person. Next we take the squared deviations from the grand mean and sum these squared deviations. The sum of squares for persons is

$$SS_p = \sum_{p=1}^{n_p} n_i (x_{p.} - x_{..})^2 \tag{5.5}$$

The mean squares for persons is obtained by dividing the sum of squares for persons by the degrees of freedom corresponding to this sum of squares, $n_p - 1$. The mean squares for persons is equal to n_i

Table 5.1 ANOVA of the crossed $p \times i$ design.

Source of Variation	Sum of Squares (SS)	Degrees of Freedom (df)	Mean Square (MS)	Expected Mean Square (EMS)
Persons (p)	SS_p	$n_p - 1$	$MS_p = SS_p/\mathrm{df}_p$	$\sigma^2_{pi,e} + n_i\sigma^2_p$
Items (i)	SS_i	$n_i - 1$	$MS_i = SS_i/\mathrm{df}_i$	$\sigma^2_{pi,e} + n_p\sigma^2_i$
Residual	$SS_{pi,e}$	$(n_p -1)(n_i - 1)$	$MS_{pi,e} = SS_{pi,e}/\mathrm{df}_{pi,e}$	$\sigma^2_{pi,e}$
Total	$\Sigma\Sigma(x_{pi} - x..)^2$			

times the variance of the mean scores and is equal to the total-score variance divided by n_i.

The sum of squares for items is obtained in a similar way. The simplest way to obtain the sum of squares for the residual is to compute the total sum of squares and to subtract the sum of squares for persons and the sum of squares for items. The complete ANOVA is summarized in Table 5.1. The rightmost column in this table gives the expected mean squares for the random-effects model. In the expected mean squares for persons, all variance components related to persons are included as well as the random error. This is due to the fact that the model is a random model. In a model with fixed effects, the interactions for a particular person would have summed to 0. In the model with random effects, the n_i interactions are a random sample from all possible interactions for the person. The coefficient of the variance component for persons is n_i. This coefficient is equal to the number of observations in which the person effect is involved.

$MS_{pi,e}$ estimates $\sigma^2_{pi,e}$. From the above table we obtain an estimate of σ^2_p:

$$\hat{\sigma}^2_p = (MS_p - MS_{pi,e})/n_i \tag{5.6}$$

The generalizability coefficient for the n_i-item test—universe-score variance divided by observed variance—is

$$E\rho^2_{\mathrm{Rel}} = \frac{\sigma^2_p}{\sigma^2_p + \sigma^2_{pi,e}/n_i} \tag{5.7}$$

also known as the stepped-up intraclass correlation. The expectation sign indicates that in this formula an approximation is given to the expected squared correlation of observed scores and universe scores. The coefficient is the generalizability counterpart of the reliability coefficient. Its size gives information on the accuracy with which comparisons between persons can be made. The coefficient concerns relative measurements, and this is denoted by *Rel* (Shavelson and Webb, 1991). The estimate of Equation 5.7 in terms of mean squares is

$$E\rho^2_{Rel} = \frac{MS_p - MS_{pi,e}}{MS_p} \tag{5.8}$$

The mean squares in Equation 5.8 can be written in terms of the total-score variance and the item variances. If we do so, we can derive that Equation 5.8 is identical to coefficient α, the coefficient known as a lower bound to reliability. This implies that in case of congeneric items, generalizability theory underestimates generalizability or reliability. The problem is due to the fact that the true-scale differences between congeneric measurements are taken up into the interaction term in the score composition (Equation 5.3).

5.3.2   The nested $i : p$ design

In the one-facet $i : p$ design, each person is presented with a different set of items. This situation is schematized in Figure 5.4. It is clear from the figure that the data matrix is incomplete.

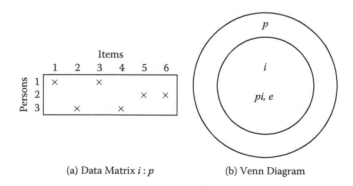

(a) Data Matrix $i : p$ (b) Venn Diagram

Figure 5.4   Data matrix and Venn diagram for the nested $i : p$ design.

Table 5.2 ANOVA of the nested $i : p$ design.

Source of Variation	Sum of Squares (SS)	Degrees of Freedom (df)	Mean Square (MS)	Expected Mean Square (EMS)
Persons (p)	SS_p	$n_p - 1$	$MS_p = SS_p/\mathrm{df}_p$	$\sigma^2_{i,pi,e} + n_i\sigma^2_p$
Residual	$SS_{i,pi,e}$	$n_p(n_i - 1)$	$MS_{i,pi,e} = SS_{i,pi,e}/\mathrm{df}_{i,pi,e}$	$\sigma^2_{i,pi,e}$
Total	$\Sigma\Sigma(x_{pi} - x..)^2$			

Only two variance components can be estimated. The ANOVA for the nested design is given in Table 5.2.

With n_i items the generalizability coefficient is

$$\rho^2 = \frac{\sigma^2_p}{\sigma^2_p + \sigma^2_{i,pi,e}/n_i} \tag{5.9}$$

which is estimated by

$$\rho^2_{\mathrm{Rel}} = \frac{MS_p - MS_{i,pi,e}}{MS_p} \tag{5.10}$$

Notice the difference between the left-hand sides of Equation 5.7 and Equation 5.9. In the nested design the ratio of variance components equals the generalizability coefficient. For more on this design see Jackson (1973).

5.4 The two-facet crossed $p \times i \times j$ design

Now two facets define our universe. Consider a universe with items and judges as facets. We have n_p persons, n_i items, and n_j judges. All persons answer all items. Each judge rates the answer of each person to each item. All combinations of n_p persons, n_i items, and n_j judges occur. We have a fully crossed $p \times i \times j$ design. The Venn diagram for this design appears in Figure 5.5. The ANOVA for this design with random effects is given in Table 5.3. Now we have a residual consisting of random error and the three-way interaction. There are three two-way interactions

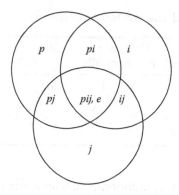

Figure 5.5 Venn diagram for the two-facet crossed $p \times i \times j$ design.

and three main effects. Again the coefficients for the variance components in the expected mean squares are equal to the number of times an effect is present in the study. For each person $n_i n_j$ observations are available; therefore, the coefficient for the variance component for persons equals $n_i n_j$.

Table 5.3 ANOVA of the crossed $p \times i \times j$ design.

Source of Variation	Degrees of Freedom (df)	Expected Mean Square (EMS)
Persons (p)	$n_p - 1$	$\sigma^2_{pij,e} + n_i\sigma^2_{pj} + n_j\sigma^2_{pi} + n_i n_j \sigma^2_p$
Items (i)	$n_i - 1$	$\sigma^2_{pij,e} + n_p\sigma^2_{ij} + n_j\sigma^2_{pi} + n_p n_j \sigma^2_i$
Judges (j)	$n_j - 1$	$\sigma^2_{pij,e} + n_p\sigma^2_{ij} + n_i\sigma^2_{pj} + n_p n_i \sigma^2_j$
Interaction pi	$(n_p - 1)(n_i - 1)$	$\sigma^2_{pij,e} + n_j\sigma^2_{pi}$
Interaction pj	$(n_p - 1)(n_j - 1)$	$\sigma^2_{pij,e} + n_i\sigma^2_{pj}$
Interaction ij	$(n_i - 1)(n_j - 1)$	$\sigma^2_{pij,e} + n_p\sigma^2_{ij}$
Residual	$(n_p - 1)(n_i - 1)(n_j - 1)$	$\sigma^2_{pij,e}$

Analogous to Formula 5.7, we obtain a generalizability coefficient:

$$E\rho^2_{Rel} = \frac{\sigma^2_p}{\sigma^2_p + \sigma^2_{pi}/n_i + \sigma^2_{pj}/n_j + \sigma^2_{pij,e}/n_i n_j} \tag{5.11}$$

When we set the expected mean squares in Table 5.3 equal to the observed mean squares, we can solve the seven equations for the seven variance components. We start in the bottom row of the table; the residual component is set equal to the observed residual mean squares. Next, we compute the variance component for the interaction between items and judges and so forth. It is possible that while doing so negative estimates of variance components are obtained. The best one can do is to compute all components using possibly negative values. After all components have been computed, we set negative values equal to zero. This is the way estimation proceeds in the simultaneous estimation procedure of some software packages. Brennan (2001) discusses the handling of negative estimates of variance components.

The generalizability estimate in terms of mean squares is

$$E\rho^2_{Rel} = \frac{MS_p - MS_{pi} - MS_{pj} + MS_{pij,e}}{MS_p} \tag{5.12}$$

Formula 5.12 can also be rewritten in terms of variances and covariances. Let us first examine the structure of the variance–covariance matrix in the crossed design. We do this with the help of Figure 5.6.

		Judge 1		Judge 2	
		Item 1	Item 2	Item 1	Item 2
Judge 1	Item 1	$\sigma^2_{1(1)}$	$\sigma_{1(1)2(1)}$	$\sigma_{1(1)1(2)}$	$\sigma_{1(1)2(2)}$
	Item 2	$\sigma_{2(1)1(1)}$	$\sigma^2_{2(1)}$	$\sigma_{2(1)1(2)}$	$\sigma_{2(1)2(2)}$
Judge 2	Item 1	$\sigma_{1(2)1(1)}$	$\sigma_{1(2)2(1)}$	$\sigma^2_{1(2)}$	$\sigma_{1(2)2(2)}$
	Item 2	$\sigma_{2(2)1(1)}$	$\sigma_{2(2)2(1)}$	$\sigma_{2(2)1(2)}$	$\sigma^2_{2(2)}$

Figure 5.6 Variances and covariances in the crossed design with 2 judges and 2 items.

Only the covariances $\sigma_{1(1)2(2)}$ and $\sigma_{1(2)2(1)}$ are "pure" covariances—covariances between different items, rated by different judges. If we denote pure covariances by $\sigma_{i(j)i'(j')}$, then we can write the following reliability coefficient (cf. Equation 4.10):

$$\rho_{XX'} = \frac{n_i^2 n_j^2 \overline{\sigma_{i(j)i'(j')}}}{\sigma_X^2} = \frac{n_i n_j}{(n_i-1)(n_j-1)}\left(1 - \frac{\sum_{i=1}^{n_i}\sigma_i^2 + \sum_{j=1}^{n_j}\sigma_j^2 - \sum_{i=1}^{n_i}\sum_{j=1}^{n_j}\sigma_{i(j)}^2}{\sigma_X^2}\right)$$

(5.13)

where σ_i^2 is the total score variance for item i, and σ_j^2 is the total score variance for judge j. The coefficient of Equation 5.13 is identical to the generalizability coefficient (Equation 5.12).

In generalizability theory, the emphasis is not so much on the estimation of reliability or, better, generalizability, as on the estimation of the variance components. The relative size of a component gives us information on the influence that this component has on measurement error. After the components have been estimated we can do a D study. We are then able to compute the generalizability coefficient for a number of items or a number of judges different from that in the G study. We are also able to estimate the effect of using another design to obtain observations. Let us restrict ourselves to the application of the crossed design. Let n_i' be an arbitrary number of items and n_j' an arbitrary number of judges, then the following formula gives the generalizability coefficient that will be obtained for these numbers of items and judges:

$$E\rho_{\text{Rel}}^2 = \frac{\sigma_p^2}{\sigma_p^2 + \sigma_{pi}^2/n_i' + \sigma_{pj}^2/n_j' + \sigma_{pij,e}^2/n_i'n_j'}$$

(5.14)

This formula is a generalization of the Spearman–Brown formula for a lengthened test. With the formula we can investigate the effect on generalizability of increasing the number of items and the effect of increasing the number of judges.

Generalizability theory is not the only possibility for reliability estimation with more than one facet. A crossed design might also be analyzed with structural equation modeling; for examples, see Blok (1985) and Werts, Breland, Grandy, and Rock (1980).

5.5 An example of a two-facet crossed $p \times i \times j$ design: The generalizability of job performance measurements

Webb, Shavelson, Kim, and Chen (1989) studied the reliability of scores of job performance of Navy machinist mates in the perspective of G theory. Three raters (supervisor, peer, and self) rated the machinist mates on four measures of job performance: a hands-on performance test, a paper-and-pencil job knowledge test, job task performance ratings, and global ratings. G theory was utilized to estimate the reliability of the measures (G study) and to improve the measurement design (D study).

Let us look at one part of the study—the ratings on the hands-on performance tests. Two examiners or raters observed each machinist on 11 tasks in the engine room. Details of the procedure are given by Webb et al. (p. 97–98).

Table 5.4 gives some results of a G study and a D study of the hands-on performance tests in terms of estimated variance components and the generalizability coefficients for relative errors. From the estimated variance components, we see that *examiner* was a negligible source of variation. This also holds for the interaction effects of persons and examiners, and tasks and examiners. It may be concluded that examiners rank machinist mates highly similarly on the hands-on performance tests. The main effect for tasks, however, is relatively

Table 5.4 Estimated variance components and generalizability coefficients for hands-on performance tests (G study and D study).

Source of Variation	Variance Components
Persons	0.00626
Examiners	0.00000
Tasks	0.00970
Persons × examiners	0.00000
Persons × tasks	0.02584
Examiners × tasks	0.00003
Residual	0.00146

Size of Design

Number of examiners	1	1	2	1
Number of tasks	1	11	11	17
G coefficient (relative)	0.19	0.72	0.72	0.80

high. This means that tasks differ in difficulty. The variance component for the interaction between persons and tasks also is relatively large. It accounts for 60% of the variability of the scores X_{pij}. Differences between persons were greater for some tasks than for others. This all has important implications for further improving the measurement of job performance using a variety of tasks. Most influential for reliability or generalizability is to introduce more tasks. From Table 5.4 it can be seen what the influence on the G coefficient would be with 17 tasks given only one examiner. Also, the generalizability for absolute decisions, a subject that will be discussed in another section, would be largely improved with an increase in the number of tasks.

How would a classical reliability study on the data be carried out? Researchers may differ on which reliability coefficient to calculate. Is a reliability coefficient as such informative in the study described in the example? The answer is no. Estimated variance components are the preferred statistics to compute (cf. *Standards*, APA et al., 1999, Chapter 2). Not only can everybody then calculate his or her own reliability or generalizability coefficient, there are also clear indications of how to improve measurement by varying the "size of the study design."

5.6   The two-facet nested $p \times (i : j)$ design

In Section 5.3 the crossed $p \times i$ design and the nested $i : p$ design were discussed. A counterpart of the crossed $p \times i \times j$ design is the partially nested design $p \times (i : j)$. The crossed and the partially nested design have been presented in Figure 5.1, with the same number of observations for each person. In the crossed design, each judge rated all answers to all items. In the nested design, each judge had another set of items at his or her disposal to judge the persons.

Apparently, the nested design is less informative than the crossed design. In the crossed design, judges can be compared; they judge the same items answered by the same persons. This comparison is not possible in the nested design. The nested design has, however, an unexpected advantage. Each person answers more items in this design. This suggests that the design might result in a higher reliability and, consequently, might be more efficient. We will demonstrate that this is the case. First, we will analyze the nested $p \times (i : j)$ design.

Let us have n_j judges. Each judge rates another set of n_i items; items are *randomly* allocated to judges. (When each judge is expert on one subject area and rates only answers to questions pertaining to that particular subject area, the universe can be regarded as nested,

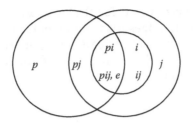

Figure 5.7 The Venn diagram for the $p \times (i : j)$ design.

and the variance components are interpreted in a slightly different way.) Each person answers $n_i n_j$ items.

We can use the Venn diagram in Figure 5.7 as a means of finding which variance components are confounded. Because p and j are crossed, their circles intersect in the diagram. The circles for p and i also intersect. The circle for i lies entirely within the circle for j, visualizing the nesting of i within j. In the diagram i and ij are found in the same segment, from which we may conclude that these effects are confounded. Also, pi is confounded with the residual pij,e. The ANOVA of the nested $p \times (i : j)$ design is given in Table 5.5.

We write the generalizability coefficient as

$$E\rho^2_{\text{Rel}} = \frac{\sigma^2_p}{\sigma^2_p + \sigma^2_{pj}/n_j + \sigma^2_{pi,pij,e}/n_i n_j} \qquad (5.15)$$

Table 5.5 ANOVA of the nested $p \times (i : j)$ design.

Source of Variation	Degrees of Freedom (df)	Expected Mean Square (EMS)
Persons (p)	$n_p - 1$	$\sigma^2_{pi,pij,e} + n_i \sigma^2_{pj} + n_i n_j \sigma^2_p$
Judges (j)	$n_j - 1$	$\sigma^2_{pi,pij,e} + n_p \sigma^2_{i,ij} + n_i \sigma^2_{pj} + n_p n_i \sigma^2_j$
Interaction pj	$(n_p - 1)(n_j - 1)$	$\sigma^2_{pi,pij,e} + n_i \sigma^2_{pj}$
Items within judges ($i : j$)	$n_j(n_i - 1)$	$\sigma^2_{pi,pij,e} + n_p \sigma^2_{i,ij}$
Residual	$(n_p - 1)n_j(n_i - 1)$	$\sigma^2_{pi,pij,e}$

The generalizability coefficient for the nested $p \times (i : j)$ design with n_i items per judge and n_j judges also can be obtained from the results of an analysis with the crossed design. The generalizability coefficient for the nested design, written in terms of the variance components from a crossed analysis, equals

$$E\rho^2_{\text{Rel}} = \frac{\sigma^2_p}{\sigma^2_p + \sigma^2_{pj}/n_j + \sigma^2_{pi}/n_i n_j + \sigma^2_{pij,e}/n_i n_j} \qquad (5.16)$$

The contribution of the variance component for the interaction persons × items to the observed variance for persons is smaller than in the crossed design. In the crossed design only n_i items are involved, in the nested design $n_i n_j$. Consequently, the denominator in Equation 5.16 is smaller than the denominator in Equation 5.11 for the crossed design with the same number of observations, and the generalizability coefficient for the nested design is higher than that for the crossed design.

5.7 Other two-facet designs

Four other types of nested designs with two facets can be distinguished:

$i \times (j : p)$
$j : (i \times p)$
$(i \times j) : p$
$j : i : p$

The $i \times (j : p)$ design is formally identical to the $p \times (i : j)$ design. In the $i \times (j : p)$ design, the persons and one of the two facets have changed places. So, also in the $i \times (j : p)$ design, five variance components can be estimated. An example of the $i \times (j : p)$ design is the design in which the responses to a set of items are judged by a group of judges, and the group of judges differs from person to person.

The $j : (i \times p)$ design and the $(i \times j) : p$ design are not equal formally. These two designs are similar in that in both designs four variance components can be estimated. Let us consider the $o : (i \times p)$ design, where o designates *occasions*. An example of the $o : (i \times p)$ design is a design where each person responds to the same tasks, but the measurement occasions differ, for persons as well as for tasks. In the $(o \times j) : p$ design, a group of judges rates the performance of a person, and each person is rated at different occasions and by a different group of judges. For example, person 1 is judged at occasions 1, 2, and 3, by judges 1 and 2; person 2 is judged at occasions 4, 5, and 6, by judges 3 and 4; and so on.

Finally, an example of a fully nested $i : j : p$ design is the situation where each person's work is judged by a different group of judges, and each judge uses another set of tasks i. In the $i : j : p$ design, only three variance components can be estimated, due to the confounding of many effects.

Cronbach et al. (1972) mention another type of design—a design in which the nested effect j in the combination $(j : i)$ occurs only once (i.e., $n_j = 1$). In this case, the notation (i,j) for "i joint with j" is used. There are two two-facet designs with "i joint with j": the $(i,j) \times p$ design and the $(i,j) : p$ design. In the $(i,j) \times p$ design, three variance components can be estimated and in the $(i,j) : p$ design, only two. If facet j is considered to have no influence on the score variation, the designs can be simplified: the $(i,j) \times p$ becomes the one-facet $i \times p$ design, and the $(i,j) : p$ design becomes the one-facet $i : p$ design.

The Venn diagrams for the four (partially) nested two-facet designs are presented in Figure 5.8.

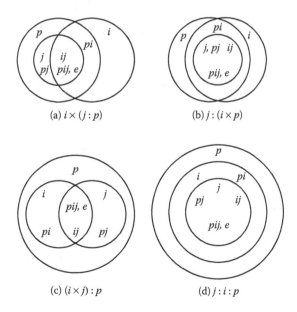

(a) $i \times (j : p)$ (b) $j : (i \times p)$

(c) $(i \times j) : p$ (d) $j : i : p$

Figure 5.8 Venn diagrams for four nested designs.

5.8 Fixed facets

The definition of measurement error and universe score depends on the extent to which one is willing to generalize over measurement conditions. Let us take the crossed design with facet items and judges. Perhaps we are not interested in generalizing over the judges in the study to a hypothetical universe of judges. Maybe the judges in the study are the only judges available for a long period of time. If we do not generalize over judges, we can redefine the effects in such a way that the interactions with judges total to zero. For example, the sum of the person × judge interactions for a particular person equals zero. An alternative way of saying this is that the average person × judge interaction for a particular person is taken up into the universe score for that person. The variance component for the person × judge interaction also becomes part of the universe score variance. The generalizability coefficient for a fixed facet *judges* can be written in terms of the initial variance components as

$$
E\rho^2_{\text{Rel}} = \frac{\left[\sigma^2_p + \sigma^2_{pj}/n_j\right]}{\left[\sigma^2_p + \sigma^2_{pj}/n_j\right] + \left(\left[\sigma^2_{pi} + \sigma^2_{pij}/n_j\right] + \sigma^2_e/n_j\right)/n_i} \tag{5.17}
$$

Generalizability is higher when a facet is fixed. The numerator in Equation 5.17 is larger than the numerator in Equation 5.11, while the denominator remains the same. This is understandable because generalization over a more limited universe is easier.

The estimated generalizability for the crossed $p \times i \times j$ design with judges fixed equals

$$
E\rho^2_{\text{Rel}} = \frac{MS_p - MS_{pi}}{MS_p} \tag{5.18}
$$

which not only looks similar to the corresponding coefficient in the crossed $p \times i$ design, but which in fact gives the same outcome as Equation 5.8. So, if judges in the crossed $p \times i \times j$ design are fixed, we can neglect the facet *judges*. We total over judges and analyze the resulting crossed $p \times i$ design. Why should we analyze the observations as a crossed $p \times i \times j$ design? An argument in this particular case to analyze the observations for the full model is that without much difficulty more information is obtained with respect to the relevant variance components.

Table 5.6 ANOVA of the nested $p \times (i : s)$ design.

Source of Variation	Expected Mean Square (*EMS*)
Persons (p)	$\sigma^2_{pi:s,e} + n_i \sigma^2_{ps} + n_i n_s \sigma^2_p$
Subtests (s)	$\sigma^2_{pi:s,e} + n_p \sigma^2_{i:s} + n_i \sigma^2_{ps} + n_p n_i \sigma^2_s$
Interaction ps	$\sigma^2_{pi:s,e} + n_i \sigma^2_{ps}$
Items within subtests ($i : s$)	$\sigma^2_{pi:s,e} + n_p \sigma^2_{i:s}$
Residual	$\sigma^2_{pi:s,e}$

Also in a nested design facets can be fixed. Let us consider an example. A test has been constructed with a number of subtests. The subtests might be tests of different aspects of the subject matter or various scales that can be distinguished. In a test on test theory, subtests might be *classical test theory* and *generalizability theory*. Multiple forms of the test can be constructed, but all tests are to be constructed with the same division into subtests. The facet *Subtests* is fixed; no generalization is sought over subtests. Items are nested within subtests and persons are crossed with items. The design is a nested $p \times (i : s)$ design. Table 5.5 has been repeated as Table 5.6. The mean squares in Table 5.6 are still specified according to the fully random design. There is a difference between both tables. In Table 5.6, the notation $i : s$ is used, instead of i,is. This is done in order to indicate that the nesting is not a result of a design decision, but the result of a nesting of the universe. Items belong to a specific subtest.

When *Subtests* is a fixed facet, the variance components must be redefined. The variance components for the model with fixed facet *Subtests* are

$$\sigma^2_{p*} = \sigma^2_p + \frac{\sigma^2_{ps}}{n_s} \tag{5.19}$$

where n_s is the number of subtests,

$$\sigma^2_{i*} = \sigma^2_{i:s} \tag{5.20}$$

and

$$\sigma^2_{pi,e^*} = \sigma^2_{pi:s,e} \qquad (5.21)$$

One might use the results in Table 5.6 to obtain a generalizability coefficient. Unfortunately, the resulting formulas are not very useful when a different number of items are allowed for each subtest or stratum. The generalizability coefficient for a test with fixed subtests or, rather, strata was derived by Rajaratnam et al. (1965). They proposed an alternative estimate of the generalizability coefficient that is also valid when the number of items varies from stratum to stratum. In the derivation of the formula, they used the total-score variance and a weighted sum of the residuals $MS_{pi,e(s)}$ for the various strata s ($s = 1,...,n_s$). The generalizability coefficient they derived is

$$\rho^2_{Rel} = \frac{\sigma^2_X - \sum_{s=1}^{n_s} n_{i(s)} MS_{pi,e(s)}}{\sigma^2_X} \qquad (5.22)$$

where $n_{i(s)}$ is the number of items in stratum s. This coefficient can be rewritten as coefficient α_s, the coefficient that was introduced in Chapter 4 (Equation 4.11).

In the analysis above, the scores on the subtests are averaged. Shavelson and Webb (1991) argue that separate analyses for the subtests should also be done. If the results differ strongly between subtests, it might be profitable to use scores on the subtests instead of, or in addition to, a total score. When scores on the subtests are obtained, we can apply multivariate generalizability theory in order to obtain estimated universe scores defined on the subtests. Cronbach et al. (1972) mention the application of the multivariate approach in connection with the estimation of universe scores for the Wechsler Performance and Verbal scales. Their results are the generalizability theory equivalent of Equation 4.12.

There are two other designs in which a fixed effect makes sense: the $i \times (j : p)$ design and the $j : (i \times p)$ design. In both designs, i can be fixed. In both designs, the variance component for the interaction between i and p is considered to be part of the universe-score variance. It makes no sense to consider a facet as fixed when this facet is nested within a random facet. Fixing both effects in a two-facet design means that one does not want to generalize; only pure measurement error is considered to contribute to the error variance in the generalizability coefficient. Generalizability cannot be estimated when both effects are fixed, because pure error variance and at least one interaction variance are confounded.

5.9 Kinds of measurement errors

Until this moment, the variance components for the main effects of the facets have not played a role in the estimation of generalizability for the crossed design. We have argued that in a crossed $p \times i$ design all persons answer the same items, so the differences between items do not play a role when persons from the same crossed study are compared to each other. In comparing the persons from the same study, we are interested in relative measurement.

The size of the item effects is relevant when we have a nested study in which each person responds to a different set of items. The size of the item effects is also relevant if we want to compare the persons in a crossed study with other persons who have taken a different test, or if we want to compare all persons to a standard of performance. In those situations, the variation in difficulty level is an interfering factor, to be regarded as part of the measurement error—we have absolute measurement errors.

So far, we used the variance for the relative measurement:

$$\sigma^2_{\text{Rel}} = \sigma^2_{pi,e}/n_i \tag{5.23}$$

With absolute errors, we have an error variance equal to

$$\sigma^2_{\text{Abs}} = \sigma^2_i/n_i + \sigma^2_{pi,e}/n_i \tag{5.24}$$

Instead of the generalizability coefficient (Equation 5.7) for relative measurement errors, we can define a coefficient for absolute measurement errors as

$$\varphi = \frac{\sigma^2_p}{\sigma^2_p + \sigma^2_i/n_i + \sigma^2_{pi,e}/n_i} \tag{5.25}$$

The coefficient is estimated by

$$\hat{\varphi} = \frac{MS_p - MS_{pi,e}}{MS_p + (MS_i - MS_{pi,e})/n_p} \tag{5.26}$$

or by the biased estimator

$$\hat{\phi} = \frac{n_i}{n_i - 1}\left(\frac{s_x^2 - \sum\limits_{i=1}^{n_i} s_i^2}{s_x^2 + n_i^2 s_c^2/(n_i - 1)} \right) \tag{5.27}$$

where s_x^2 is the total-score variance (with denominator n_p), s_i^2 the variance of condition i (with denominator n_p), and s_c^2 the variance of the condition means (with denominator n_i) (Rajaratnam, 1960).

We denote coefficient φ the *index of dependability* following the suggestion by Brennan and Kane (1977). This coefficient cannot be regarded as a correlation except in the nested design. In the nested $i : p$ design, relative errors and absolute errors are identical. For a comparison of the designs and coefficients, see Exhibit 5.1.

Exhibit 5.1   The one-facet crossed $p \times i$ design and the nested $i : p$ design

DESIGN $p \times i$

Estimated variance components:

$$\hat{\sigma}_p^2, \hat{\sigma}_i^2, \hat{\sigma}_{pi,e}^2$$

Generalizability coefficient given n_i items (relative decisions):

$$E\hat{\rho}^2 = \frac{\hat{\sigma}_p^2}{\hat{\sigma}_p^2 + \hat{\sigma}_{pi,e}^2/n_i}$$

Index of dependability given n_i items (absolute decisions):

$$\phi = \frac{\hat{\sigma}^2_p}{\hat{\sigma}^2_p + \hat{\sigma}^2_i/n_i + \hat{\sigma}^2_{pi,e}/n_i}$$

DESIGN $i : p$

Estimated variance components:

$$\hat{\sigma}^2_p, \hat{\sigma}^2_{i,pi,e} \left(= \text{estimate of } \sigma^2_i + \sigma^2_{pi,e}\right)$$

Generalizability coefficient given n_i items:

$$\hat{\rho}^2 = \frac{\hat{\sigma}^2_p}{\hat{\sigma}^2_p + \hat{\sigma}^2_{i,pi,e}/n_i}$$

In this nested design there is no difference between absolute and relative decisions.

The index of dependability for the crossed design estimates the generalizability for the nested $i : p$ design.

The index of dependability has a lower value than the coefficient for relative measurement errors because more variance components contribute to the absolute error. The variance component for items in the crossed $p \times i$ design is based on the assumption of random sampling of items. If items vary in difficulty level, we probably are not prepared to sample items randomly from the universe. We will stratify the universe and use a stratified random sampling scheme instead. Within strata, items will vary less in difficulty. When items are sampled from a stratified universe, the absolute error variance is overestimated with Equation 5.24.

The relative and absolute errors might be viewed as two extreme possibilities to think about when discussing errors and decision making. When using absolute errors, we implicitly assume that the scores are not corrected for measurement bias. We might, for example, use a test known to be relatively difficult without correcting the test scores. The alternative is, of course, to use score corrections wherever possible. In many situations, enough knowledge with respect to the relative difficulty of alternative tests is available in order to correct scores at least partially.

Let us take the following model:

$$
\begin{array}{lll}
X_{pi} = & \mu & \text{Grand mean} \\
& + \mu_p - \mu & \text{Person effect} \qquad (5.28) \\
& + \mu_i - \mu & \text{Condition effect} \\
& + X_{pi} - \mu_p - \mu_i + \mu & \text{Residual}
\end{array}
$$

The score x_{pi} obtained under condition i can be corrected for the condition effect $\mu_i - \mu$, giving

$$
x'_{pi} = x_{pi} - (\mu_i - \mu) \qquad (5.29)
$$

Suppose that two tests X and Y have been administered to two large random samples from the population. Then the condition means μ_i, the population means of the two tests, are known and scores on both tests can be corrected by taking deviation scores $x_{pi} - \mu_i$. So, test scores are made comparable with a relative measurement approach—an approach that makes use of the results obtained by a group of examinees on a test. In Chapter 11 the conditions under which scores on different tests can be made equivalent are discussed more extensively.

The correction in Equation 5.29 is an ideal, however. In practice, the effects $\mu_i - \mu$ needed for the correction are imperfectly estimated. If the number of persons tested under condition i is very small, we have little information on the size of the value μ_i. Then it is sensible not to correct, and the absolute error is the relevant error for absolute decision making. With a larger number of persons, the condition mean $x_{.i}$ contains useful information on the value of μ_i. But the influence of measurement error may still be so large that we should not estimate μ_i by the mean condition score. However, if some information on the variation of condition effects is available, one might estimate the condition effect μ_i using a Kelley formula. Such a procedure was suggested by De Gruijter and Van der Kamp (1991), based on work by Lindley (1971) on what is to be regarded as a very simple multilevel model (Snijders and Bosker, 1999). The equation for the estimation of condition mean μ_i based on the results of n_i persons is

$$
\hat{\mu}_i = \hat{\rho}_i^2 x_{.i} + \left(1 - \hat{\rho}_i^2\right)\hat{\mu} \qquad (5.30)
$$

where $x_{.i}$ is the observed mean for condition i,

$$\hat{\mu} = \frac{\sum_{j=1}^{m} w_j x_{.j}}{\sum_{j=1}^{m} w_j} \tag{5.31}$$

$$\hat{\rho}_i^2 = \frac{\hat{\sigma}_b^2}{\hat{\sigma}_b^2 + \hat{\sigma}_{p,res}^2 / n_i} \tag{5.32}$$

and

$$w_i = \frac{1}{\hat{\sigma}_b^2 + \hat{\sigma}_{p,res}^2 / n_i} \tag{5.33}$$

An estimate of the within-condition variance is

$$\hat{\sigma}_{p,res}^2 = \frac{\sum_{j=1}^{m} \sum_{p=1}^{n_j} (x_{pj} - x_{.j})^2}{\sum_{j=1}^{m} n_j - m} \tag{5.34}$$

The variance component for conditions can be estimated as

$$\hat{\sigma}_b^2 = \frac{\sum_{j=1}^{m} (x_{.j} - \bar{x})^2}{m-1} - \frac{\left(\sum_{j=1}^{m} \frac{1}{n_j}\right)}{m} \hat{\sigma}_{p,res}^2 \tag{5.35}$$

with

$$\bar{x} = \frac{\sum_{j=1}^{m} x_{.j}}{m} \tag{5.36}$$

(Jackson, 1973).

The formulas allow for conditions with different numbers of persons. The weight of an observed condition mean for the computation of the estimate of the universe mean is relatively high when the number of persons who have been measured under that condition is large. From Equation 5.30 and Equation 5.32, we see that in this case the observed mean score for the condition is close to the true condition mean.

Now scores can be corrected for the condition effect. But, for an optimal estimate of a person's universe score, Kelley's formula can be used. So, an estimate of the universe score is

$$\hat{\mu}_p = \rho^2_{\text{Rel}}(x_{pi} - \hat{\mu}_{.i}) + \hat{\mu} \qquad (5.37)$$

where ρ^2_{Rel} is the reliability of condition i. The error associated with the procedure obviously differs from the absolute error. It also differs from the relative error unless the condition mean and universe mean are perfectly estimated.

The estimation formula can be rewritten as

$$\hat{\mu}_p = \hat{\mu} + \rho^2_{\text{Rel}}(x_{pi} - x_{.i}) + \alpha(x_{.i} - \hat{\mu}) \qquad (5.38)$$

where α is the product of the generalizability coefficient and $1 - \rho^2_i$. Jarjoura (1983; see also Longford, 1994) discusses the optimal estimation of universe scores on the basis of a n-item test without taking the intermediate step of estimating the condition mean μ_i. When items are randomly selected for test forms, Equation 5.38 is identical to his formula (39).

Related to the sampling approach in generalizability theory is *matrix sampling* (for an overview see Lord and Novick, 1968, and Sirotnik and Wellington, 1977). In matrix sampling from a population of persons and a one-faceted universe, the restrictions in the choice among designs used in generalizability theory can be dropped. In large-scale testing programs, program evaluation and the measurement of group performance matrix sampling is applied. Here the term *matrix sampling* refers to a measurement format in which a large set of test items is organized into relatively short item sets, each of which is randomly assigned to a subsample of test takers, so avoiding the administration of all items to all examinees (*Standards*, APA et al., 1999, p. 178).

Other designs than the crossed design and the nested design allow us to efficiently estimate condition effects. Figure 5.9 shows a design

	Group 1	Group 2	Group 3	Group 4
Judge 1	×	×		
Judge 2		×	×	
Judge 3			×	×
Judge 4	×			×

Figure 5.9 A judgmental design with overlap between judges.

where judges overlap. So, judge 1 and judge 2 have group 2 in common, but only judge 1 rates the persons from group 1, and only judge 2 rates persons in group 3. Through the overlap, all judges are connected and the relative leniency of judges can be estimated. With a design like the design in Figure 5.9, we have already left the domain of generalizability theory. For an overview of methods to estimate judgmental effects, see the contribution by Braun (1988).

5.10 Conditional error variance

Another important issue in summarizing reliability data and errors of measurement, not yet discussed in this chapter, is the reporting of conditional error variance (see *Standards*, APA et al., 1999, p. 27). Generalizability theory allows for a conditional standard error of measurement and a conditional error variance (i.e., the error variance varies over particular levels of scores). It is more likely that the error variance is not constant than that it is constant. Brennan (1998) gives information on the estimation of the conditional error variance, for absolute as well as for relative measurements.

Let us consider the single-facet design. The estimate of the absolute error variance for person p is

$$\hat{\sigma}^2_{\text{Abs}(p)} = \frac{\sum_{i=1}^{n_i}(x_{pi} - x_{p.})^2}{n_i(n_i - 1)} \tag{5.39}$$

This estimate of the conditional error variance is valid for the crossed $p \times i$ design as well as the nested $i : p$ design. When the items are scored dichotomously, the relevant model is the binomial model, and Equation 5.39 can be simplified to Equation 6.3 divided by the square of the number of items.

The estimation of the conditional relative error variance for the $p \times i$ design by Brennan is based on work by Jarjoura (1986). The conditional relative error variance, the variance of δ_p, can be written as

$$
\begin{aligned}
\sigma^2_{\text{Rel}(p)} &= \text{var}[(X_{pI} - \mu_I) - (\mu_p - \mu) \mid p] \\
&= \sigma^2_{\text{Abs}(p)} + \sigma^2_i / n_i - 2\text{cov}(X_{pI} - \mu_p, \mu_I - \mu \mid p) \qquad (5.40) \\
&= \sigma^2_{\text{Abs}(p)} - \sigma^2_i / n_i - 2\text{cov}(\mu_i - \mu, X_{pi} - \mu_p - \mu_i + \mu \mid p)/n_i
\end{aligned}
$$

where X_{pI} and μ_I are means over n_i-item sets. Averaged over persons, the latter covariance term is zero, and the relation between absolute errors and relative errors as given in Equation 5.23 and Equation 5.24 is obtained. The conditional relative error variance is estimated as

$$
\hat{\sigma}^2_{\text{Rel}(p)} = \frac{n_p + 1}{n_p - 1} \hat{\sigma}^2_{\text{Abs}(p)} + \hat{\sigma}^2_i / n_i - 2 \left(\frac{n_p}{n_p - 1} \right) \frac{\text{cov}(x_{pi}, x_{.i} \mid p)}{n_i} \qquad (5.41)
$$

Brennan considers other designs as well. One of these designs is the design in which a table of specifications is used for the stratification of items. The estimated conditional absolute error variance for the stratified design is a generalization of the conditional error variance derived by Feldt (1984) for dichotomous items and mentioned in the next chapter.

5.11 Concluding remarks

After more than a century testing, a broadening of measurement, and at the same time a development in sophistication, is visible. In addition to traditional measurement, so-called assessments become more important (cf. the *Standards*, APA et al., 1999). A broadening of applications in practice goes hand in hand with the development of more sophisticated statistical models for test scores obtained in all kinds of assessments and also in program evaluation. G theory can be used in the analysis of sources of variation of assessments. Examples of the use of G theory in the assessment of student achievement and school effectiveness have been given by Cronbach et al. (1997). Looking back at more than four decades of G theory, however, one must come to the conclusion that G theory is underutilized, in spite of recent work in the field.

Applications and developments in specific theoretical areas are the analyses of quantitative behavioral observational data (Suen and Ary, 1989; see also Rogosa and Ghandour, 1991). G theory in the context of repeated measurements is only incidentally mentioned (Shavelson Webb, and Rowley, 1989). Every now and then authors are propagating G theory for designing, assessing, and improving the dependability of measurement procedures (e.g., Marcoulides, 1999): hardly to any avail. There are rays of hope for the future, however. The relevance of the use of G theory and G coefficients have found ample place in the 1999 *Standards for Educational and Psychological Testing* (APA et al., 1999, Chapter 2).

Exercises

5.1 We have the following table:

Person	Item 1 2 3 4 5 6 7 8 9 10
1	1 1 0 1 1 0 0 1 1 0
2	1 0 1 0 1 1 0 0 0 0
3	0 0 0 1 1 1 0 1 0 1
4	1 1 0 1 0 0 0 0 0 1
5	1 1 1 1 0 1 0 0 0 0
6	1 1 0 0 0 0 0 1 0 1
7	1 0 1 0 1 0 0 0 1 1
8	0 0 0 0 1 1 0 0 0 0
9	1 1 1 1 1 0 1 1 1 0
10	1 1 1 1 1 1 1 1 0 0
11	1 1 1 1 1 0 0 1 0 1
12	1 1 1 0 0 1 1 1 1 1
13	1 1 1 1 1 0 1 1 0 0
14	0 1 1 1 0 0 1 0 1 1
15	1 0 1 1 1 1 1 1 1 0
16	1 0 0 1 0 1 1 0 1 0
17	1 1 0 1 1 0 1 1 1 0
18	1 1 1 1 1 1 0 1 1 1
19	1 1 1 1 1 1 1 1 0 1
20	1 1 1 1 1 1 1 1 0 1

Compute the item variances and the variance of total scores. Next, compute coefficient α.

Compute the mean squares for items, persons, and interaction. Compute the variance components and discuss the implications of the values of these components. Finally, compute the generalizability coefficient. Use statistical software if you want to.

5.2 A test consisting of 15 open-answer items is given to 500 examinees. The responses are judged by four judges in a completely crossed design. The mean squares from an ANOVA are given in the table below. Compute the variance components and the generalizability coefficient for 15 items and 4 judges.

Source of Variation	Mean Square (MS)	Source of Variation	Mean Square (MS)
Persons (p)	17.30	pj	0.80
Items (i)	1051.65	ij	45.65
Judges (j)	420.80	pij,e	0.65
pi	6.65		

5.3 Compute the generalizability coefficient for (a) 30 items and 4 judges and (b) 60 items and 2 judges, using the estimated variance components from Exercise 5.2.

5.4 The following table gives the expected mean squares for the nested $j : (i \times p)$ design. Give the coefficients of the variance components in terms of n_p, n_i, and n_j.

$$EMS \text{ of the Nested } j : (i \times p) \text{ Design}$$

EMS_p	$\sigma^2_{j,pj,ij,pij,e} + a\sigma^2_{pi} + b\sigma^2_p$
EMS_i	$\sigma^2_{j,pj,ij,pij,e} + c\sigma^2_{pi} + d\sigma^2_i$
EMS_{pi}	$\sigma^2_{j,pj,ij,pij,e} + e\sigma^2_{pi}$
$EMS_{j,pj,ij,pij,e}$	$\sigma^2_{j,pj,ij,pij,e}$

5.5 Derive the formula for the correlation between two judges who both judge the responses to n_i items. Use the notation of the variance components from generalizability theory (cf. Maxwell and Pilliner, 1968).

5.6 Derive the formulas for the relative and absolute error variance for the crossed $p \times i \times j$ design.

5.7 Three judges rated 50 examinees each. The variances of the ratings are practically equal for all three judges. The pooled within-judges variance equals 100.0. The judges have different means. Judge 1 has a mean equal to 32.0, judge 2 has a mean equal to 35.0, and judge 3 has a mean equal to 38.0. Is a correction for the difference in leniency indicated? If so, how should we correct the scores?

CHAPTER 6

Models for Dichotomous Items

6.1 Introduction

The simplest items in achievement testing have only two different outcomes, correct and incorrect. These items are dichotomous. If an examinee does not answer an item, we evaluate the nonresponse as incorrect. A correct answer can be assigned a score 1, and an incorrect answer can be assigned a score 0 (see Exhibit 6.1). Dichotomous items are frequently used in tests. For example, achievement, aptitude, and intelligence tests with multiple-choice items are frequently scored dichotomously. For dichotomous and dichotomized items, test models have been developed to account for the scores of persons on such tests.

Exhibit 6.1 On the existence of dichotomous items

Dichotomous items as such do not exist, dichotomous scoring does. The responses to items do not fall naturally into two categories, "correct" and "incorrect." It takes a decision to code nonresponse and incorrect response(s) all in the same category.

In tests with multiple-choice items, sometimes a scoring formula is used in order to suppress pure guessing. The possible scores are 1 (correct), 0 (omit, not reached), and $1/(k-1)$, where k is the number of response options. When a test is not speeded, a "deterrent" against guessing is not likely to be very effective: a person should always respond to an item if he or she has some partial knowledge.

Items also can be weighted. A correct answer on one item might, for example, result in a score of one point, whereas a correct answer to another item might result in two score-points. Empirical weighting of dichotomous items will be discussed in connection with maximum likelihood estimation of person parameters (Chapter 9). Formula scoring implicitly weights two-choice items heavier than four-choice items, even though two-choice items are less accurate in the lower score range than four-choice items.

The first model to be discussed (in Section 6.2) is the binomial model. This model is relevant for the nested $i : p$ design. It is also the adequate model if items are psychometrically exchangeable. In Section 6.3 the generalized binomial model for items with varying difficulties is introduced. The generalized binomial model is the unidimensional model for dichotomous items within the context of true-score test theory. This model spans the bridge to item response models, which are discussed in Chapter 9. Section 6.4 relates the generalized binomial model to the item response models. In Section 6.5 the relevance of item statistics for item analysis and item selection is clarified.

6.2 The binomial model

If we throw an unbiased dice, the probability of obtaining the outcome five or six equals one third. When we throw the dice again, the probability again equals one third. The probability of having x times the outcome five or six in n throws is given by the binomial distribution with parameter $\zeta = 1/3$:

$$f(x \mid \zeta) = \binom{n}{x} \zeta^x (1-\zeta)^{n-x} \tag{6.1}$$

where

$$\binom{n}{x} = \frac{n!}{(n-x)!x!}$$

is the binomial coefficient with $n! = n(n-1)...1$.

In this section we will develop the binomial model, assuming the existence of a large item pool. The items are assumed to be independent: the correct answer to one item does not give away the correct response to another item. We randomly select one item from the item pool and ask a person to answer this item. The probability that this person answers a randomly selected item correctly is called his or her domain score ζ. In other words, when we repeat the testing procedure, the expected value of the proportion correct answers equals ζ.

Let us administer not one, but n randomly selected items. The probability of a correct answer is equal to ζ for each of the randomly selected items. The probability of exactly x items correct out of n is given by the binomial model presented in Formula 6.1 (actually, the binomial model is an approximation for n smaller than infinity if we use item selection without replacement). With a large number of repeated selections of n-item tests, the empirical distribution of the number correct will approximate the distribution defined by Equation 6.1.

The result of our exercise is that we have a strong true-score model with respect to the distribution of observed scores (and errors) on the basis of a few weak assumptions. The model is called strong because the error distribution given the domain score is known. There are no assumptions besides the assumption of a large item pool, and the random selection of items from this pool. It is possible for a person to know some of the items from the pool and to answer those items correctly. He or she may not know the correct answer to other items and guess correctly or not when answering these items. It might even be possible for the test administrator to know which items will be answered correctly and which will not be answered correctly. To illustrate this, suppose that a person has to respond to items on addition and subtraction. All addition items are correctly answered and none of the subtraction items. If the next item is presented and this item turns out to be an addition item, we assume that the person will answer this item correctly. Nevertheless, whether we have some information or not, over replications of n-item tests, the distribution of total score will be the binomial distribution.

Now let us consider the situation of a large item pool with more persons. If we give these persons the same selection of n items, it is unlikely that the binomial model holds. From the responses, it will become clear that the items have different psychometric characteristics. For one thing, they are likely to differ in difficulty level.

When more persons are tested, the binomial model still holds if for every person a separate random selection of items from the item pool is drawn. In terms of generalizability theory, we have a nested $i : p$ design.

The binomial model has been popular in educational testing (Hambleton and Novick, 1973). In educational testing, frequently a large domain of real or hypothetical items can be constructed and a test can be viewed as a random item selection from this item pool. The purpose of testing is to obtain an estimate of the domain score (universe score in terms of generalizability theory). Relevant questions are to which extent the person has achieved mastery of the domain, and whether the amount of mastery is enough to pass the person on the examination. In terms of generalizability theory, one is interested in absolute measurement.

An alternative to random selection of items is using a stratified sampling scheme. In relatively heterogeneous item domains, we are likely to prefer this sampling scheme. In a relatively homogeneous item domain, we might actually be prepared to select items randomly from an item pool. We will elaborate this latter possibility.

6.2.1   The binomial model in a homogeneous item domain

In the binomial model, the variance of measurement errors given test length n and domain score ζ, which is the variance of observed scores given n and ζ, equals

$$\sigma^2_{X|\zeta} = n\zeta(1-\zeta) \tag{6.2}$$

With an n-item test, the true score of person p is $\tau_p = n\zeta_p$. However, in this case, it is more convenient to keep using the true-proportion correct scale ζ. An application of the binomial model with the observed-score variance (Equation 6.2) is given in Exhibit 6.2.

Exhibit 6.2   Minimum test length

Consider the following problem. We have an ability level ζ_h that is considered as a definitely high level and another ability level ζ_l that is low. We want to classify an examinee as a high-ability examinee when $x \geq x_0$ and as low otherwise. We want to have an error probability $P(x < x_0 | \zeta_h) \leq \alpha$ for a specified high ability ζ_h. We also want to have an error probability $P(x \geq x_0 | \zeta_l) \leq \beta$ for a specified low ability ζ_l. How many items are needed to achieve the specified accuracy, and for which cut score x_0? We will discuss the simpler problem with $\beta = \alpha$.

The minimum test length is the smallest number of items n for which

$$\min_{x_0} \{\max[P(x < x_0 | \zeta_h), P(x \geq x_0 | \zeta_l)]\} \leq \alpha$$

When n is not too small and the ability ζ not too extreme, the distribution of x can be approximated by a normal distribution with mean ζ and standard deviation $n^{1/2}\sigma_\zeta = n^{1/2}[\zeta(1 - \zeta)]^{1/2}$. Let z_α be the z-score corresponding to the cumulative probability α in the normal distribution. Then x_0 and n can be obtained from the equations

$$\frac{x_0 - n\zeta_h}{\sqrt{n}\sigma_{\zeta_h}} = z_\alpha$$

and

$$\frac{n\zeta_l - x_0}{\sqrt{n}\sigma_{\zeta_l}} = z_\alpha$$

The minimum test length is

$$n = z_\alpha^2 \frac{(\sigma_{\zeta_h} + \sigma_{\zeta_l})^2}{(\zeta_h - \zeta_l)^2}$$

and the corresponding cut score

$$x_0 = n \frac{\sigma_{\zeta_h}\zeta_l + \sigma_{\zeta_l}\zeta_h}{\sigma_{\zeta_h} + \sigma_{\zeta_l}}$$

Birnbaum (1968, pp. 448–449) and Fhanér (1974) give a more general treatment of the subject. Unfortunately, the normal approximation does not always give the correct result because the minimum number of items tends to be underestimated. Part of the problem is that x_0 in the approximation is a continuous variable. For better results, the cut score should take on an integer value minus a continuity correction equal to one half. Wilcox (1976) demonstrated that an exact solution for the binomial model is feasible. In Chapter 10, another solution to the problem of minimum test length is discussed, within the framework of IRT.

The error variance of person p can be estimated from the number correct score x_p as

$$\hat{\sigma}_{E(p)}^2 = \hat{\sigma}_{X(p)}^2 = n \left[\frac{x_p - n(x_p/n)^2}{n-1} \right] = \frac{x_p(n - x_p)}{n-1} \tag{6.3}$$

The error variance is small for domain scores close to 0 and 1, and high for domain scores close to one half. It is clear that the assumption of an error variance independent of the true score level is untenable. The estimated conditional standard error of measurement on the proportion correct scale—the square root of Equation 6.3 divided by n—can be used to construct a confidence interval for ζ_p. Due to the fact that the binomial errors are asymmetrically distributed around ζ, and that the size of the variance varies with ζ, the construction of a confidence interval for ζ unfortunately is not straightforward (see Pearson and Hartley, 1970). For not too extreme proportions correct $\bar{x}_p = x_p/n$ and for not too small test sizes n a normal distribution can be used for the computation of a confidence interval around $\bar{x}_p$.

There is a second reason not to trust a confidence interval based on the observed proportion correct blindly. When we are dealing with a population of persons, such a confidence interval may well be misleading. We have to take the population distribution into account in the construction of such an interval. For a comparable situation, we refer to the discussion around the Kelley formula in Chapter 3.

What are the characteristics of the procedure with randomly selected n-item tests? How do we express reliability for the procedure in terms of the ratio of true-score variance and observed-score variance for a particular population of persons? Let us first estimate the average error variance. Using Equation 6.3, we can estimate the error variance related to observed scores X through averaging the estimated error variances for all persons. We obtain

$$\sigma_E^2 = \frac{1}{N} \sum_{p=1}^{N} \hat{\sigma}_{X(p)}^2 = \frac{1}{n-1}\left[n^2 \mu_{\bar{x}}(1-\mu_{\bar{x}}) - \sigma_X^2 \right] \qquad (6.4)$$

if, in the computation of the observed-score variance, the numerator is divided by the number of persons N instead of the usual $N-1$. In the above formula, $\mu_{\bar{x}}$ equals the proportion correct averaged over persons. This results in reliability coefficient:

$$\mathrm{KR}_{21} = \frac{n}{n-1}\left(1 - \frac{n\mu_{\bar{x}}(1-\mu_{\bar{x}})}{\sigma_X^2} \right) \qquad (6.5)$$

This coefficient is known as the Kuder–Richardson Formula 21 (Kuder and Richardson, 1937), in a crossed design a lower lower bound to reliability than KR20 (coefficient α). Here, in the nested design, the formula does not give a lower bound but is exact, apart from sampling fluctuations.

The Kelley formula for the estimation of the domain score is given by

$$\hat{\zeta}_p = \mathrm{KR}_{21}\bar{x}_p + (1 - \mathrm{KR}_{21})\mu_{\bar{x}} \qquad (6.6)$$

The regression of domain scores on observed scores or proportions is linear if the population distribution is given by a beta distribution (see Novick and Jackson, 1974). We have a linear regression with unequal error variances. In Chapter 3, linear regression of true scores on observed scores was obtained for equal error variances. If the domain scores have a beta distribution, we not only have an exact point estimate of ζ_p (Formula 6.6), but the complete posterior distribution (see Exhibit 6.3).

Exhibit 6.3   The beta–binomial complex

The beta distribution for domain scores is defined by

$$f(\zeta) \propto \zeta^{a-1}(1-\zeta)^{b-1}, \quad \text{with} \quad a, b > 0$$

Let us assume that the population distribution of domain scores is the beta distribution with parameters a and b. A person from the population answers x items from an n-item test correctly. The probability of x correct out of n for a particular value of ζ is given by

$$f(x \mid \zeta) = \binom{n}{x} \zeta^x (1-\zeta)^{n-x}$$

Notice the similarity of the beta distribution and the binomial distribution. We can derive that the posterior distribution of ζ given the test score is

$$f(\zeta \mid x) \propto \zeta^{a+x-1}(1-\zeta)^{b+n-x-1}$$

which is a beta distribution as well. A confidence interval for ζ given the observed score can be obtained; in the literature this kind of confidence interval has been designated a *credibility* interval or a *tolerance* interval.

In the figure, the distribution with the larger variation is the beta distribution with $a = 13$ and $b = 10$; its mean equals 0.57. A person answers 16 out of 20 items correctly; the proportion correct is 0.8. The more peaked distribution gives the posterior distribution of ζ given the score on the 20-item test. Its mean equals 0.67.

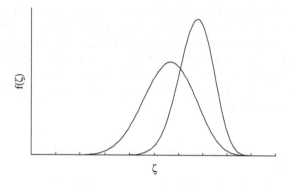

The beta distribution is also used for the construction of the exact "classical" confidence interval for ζ (Pearson and Hartley, 1970).

The nested design in which for each examinee a different random sample of items is selected, is easily implemented on the computer. With computerized testing, it is also easy to adapt the test length. If, after the administration of a number of items, the estimate of the domain score is accurate enough, testing can be stopped. A very simple stopping rule was suggested by Wilcox (1981). Wilcox assumed that there is a test procedure with a fixed test length of n items. An examinee passes the test if at least n_c items are answered correctly. This procedure can be adapted as follows:

 Stop after n_c correct responses
 Stop after $n - n_c + 1$ incorrect responses

With this procedure, test length can be much shorter than n for most examinees. The flexibility of test length is not the only characteristic of the suggested procedure, however. The procedure also assumes that each presented item has been answered. It is not possible to skip an item temporarily or to change the response to an item. Another adaptive procedure, a procedure with an optimal selection of items instead of random sampling, is discussed in Chapter 10.

6.2.2 The binomial model in a heterogeneous item domain

In a large heterogeneous item bank, the procedure for estimating the domain score, error variance, and reliability is as follows. Instead of sampling items randomly from this item bank, we randomly select items from various strata. With q strata, we randomly select n_i items from stratum i. The domain score of interest is then given by

$$\zeta_{\cdot} = \sum_{i=1}^{q} n_i \zeta_i / n \qquad (6.7)$$

with

$$n = \sum_{i=1}^{q} n_i$$

that is, the domain score $\varsigma.$ is a weighted average of the domain scores for the various strata (for a more general approach, see Jarjoura and Brennan, 1982). This domain score generally differs from the domain score in Equation 6.1. To illustrate the point, assume that the strata differ in average item difficulty. Also assume that for all strata the same number of items n_i is selected. When the strata sizes are equal, the domain score from Equation 6.7 equals the domain score under random sampling. The sizes of the strata are arbitrary, however. Some strata might contain more items than other strata (e.g., it might be easier to construct many items for some strata than for other strata). When strata differ in size, the domain score based on stratified sampling can deviate from the domain score under random sampling. Under these circumstances an analysis based on the stratified sampling plan is indicated.

The error variance in the stratified sampling approach equals

$$\sigma^2_{X|\zeta.} = \sum_{i=1}^{q} n_i \zeta_i (1 - \zeta_i) \qquad (6.8)$$

which is generally smaller than the variance that we obtain under random sampling. The estimated error variance for person p equals

$$s^2_{E(p)} = \sum_{i=1}^{q} \frac{x_{pi}(n_i - x_{pi})}{n_i - 1} \tag{6.9}$$

(Feldt, 1984). The relevant reliability coefficient is the stratified version of KR21:

$$\text{KR}_{21(s)} = \frac{\sum_{i=1}^{q} \text{KR}_{21(i)} \sigma^2_{Y_i} + \sum_{i=1}^{q} \sum_{j \neq i}^{q} \sigma_{Y_i Y_j}}{\sigma^2_X} \tag{6.10}$$

where KR21(i) is the reliability estimate for the subtest of stratum i, and Y_i designates subtest i. We should keep in mind here that each subtest contains different items for different examinees.

6.3 The generalized binomial model

We start again with an n-item test, and this time the n-item test is presented to a group of persons. Assuming that the number of items n and the number of persons N are relatively large, we are going to do some computations. We compute the correlation of an item, say item i, with the other items (see Section 6.5), and this correlation is positive. We also compute the observed proportion correct for this item within each score group on the test. Next we plot these proportions against the test scores x. The proportion correct increases with increasing test score x. The result will look like the plot in Figure 6.1. We can do the same thing for a second item, item j. It may turn out that items i and j are practically uncorrelated within each score group. We then conclude that the answers to these items are determined by one common factor (if this is the case, one actually should expect a slightly negative correlation between the two items in each score group, for the scores on the items must add to x in score group x); see Stout (1987) for a nonparametric test of unidimensionality. The common factor or latent trait score is represented by the true score on the test,

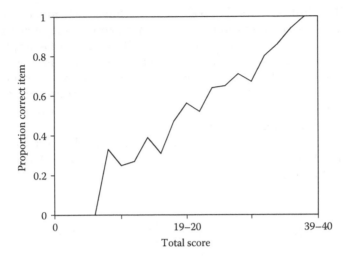

Figure 6.1 Item-test regression as might be obtained in practice.

and for a long test this true score may be reasonably well approximated by the observed score.

Again we use the true proportion correct on the test and denote this proportion correct by ζ. The proposition that there is one factor underlying the responses to the test items can be formalized as follows:

- The probability of a correct answer on item i is $P_i(\zeta)$.
- The true score on the proportion scale is $\zeta = n^{-1}\Sigma\,P_i(\zeta)$.
- Given the true score ζ the responses to items are independent. This is the property of local independence.

For two items i and j, local independence means

$$P(X_i = x_i, X_j = x_j \mid \zeta) = P(X_i = x_i \mid \zeta)P(X_j = x_j \mid \zeta)$$

$$= P_i(\zeta)^{x_i}[1-P_i(\zeta)]^{1-x_i}P_j(\zeta)^{x_j}[1-P_j(\zeta)]^{1-x_j}$$

(6.11)

where x_i equals 1 for a correct answer on item i and 0 otherwise. Formula 6.11 is shorthand for: the probability on items i and j correct is equal to the probability that i is correct times the probability that j is correct, and so forth.

Tacitly, but inevitably, we seem to have introduced a strong assumption concerning the process of answering items. From the idea that responses are locally independent, it seems to be implied that answering the items is probabilistic. This conclusion is, however, not so inevitable as it appears. Whether or not the answer process is probabilistic can only be verified in a replication study with the same test (cf. the confounding of interaction and error in Chapter 5). For reasons of convenience, we will speak of probabilities.

The model introduced above is the generalized binomial test model (Lord and Novick, 1968). The error variance given ζ defined on the scale of total scores is

$$\sigma^2_{X|\zeta} = \sum_{i=1}^{n} P_i(\zeta)[1 - P_i(\zeta)] = n\zeta(1 - \zeta) - \sum_{i=1}^{n} [P_i(\zeta) - \zeta]^2 = n\zeta(1 - \zeta) - n\sigma^2_{P|\zeta}$$

$$(6.12)$$

If the item difficulties in the generalized binomial model differ slightly for each level of ζ, the generalized binomial model can be approximated well by the binomial model. This is clear from Equation 6.12. With small differences between items, the rightmost factor in Equation 6.12 can be dropped. The more items differ with respect to difficulty, leading to a larger item variance given ζ, the smaller the error variance in the generalized binomial model is relative to the error variance of the binomial model. Does this mean that for accurate testing tests should be used with spread item difficulties? This question is not easy to answer because a different choice of items results in another true-score scale. Actually, later in this chapter it is argued that the answer should be "no" in most cases.

The error variance in the generalized binomial model varies strongly with true score. Can a reasonable estimate of error variance (Equation 6.12) be obtained for various levels of ζ? For extreme values of ζ (values close to 0 or 1) the value of Equation 6.12 is close to 0. It seems acceptable to approximate Equation 6.12 by

$$\sigma^2_{X|\zeta} \approx nk\zeta(1 - \zeta) \qquad (6.13)$$

with $0 \leq k \leq 1$. Keats (1957) proposed to choose the factor k so as to be able to reproduce the reliability coefficient $r_{XX'}$ that has been

obtained for the test. In this case, the estimate of the error variance
of person p equals

$$\hat{\sigma}^2_{E(p)} = kx_p(n - x_p)/(n-1) \qquad (6.14)$$

with

$$k = \frac{1 - r_{XX'}}{1 - KR_{21}} \qquad (6.15)$$

Feldt, Steffen, and Gupta (1985) compared various methods for the
estimation of the variance of measurement errors as a function of true
score, including the method proposed by Keats. We will discuss one of
the other methods in the next section. Another discussion of conditional
standard errors of measurement and conditional error variances can
be found in Lee, Brennan, and Kolen (2000).

6.4 The generalized binomial model and item response models

The generalized binomial model in Equation 6.11 is a general one-
factor model for dichotomous items. The probability of a correct answer
to an item increases as a function of true score in a way that is not
specified. True score is a function of the items and, therefore, is arbitrary. If we would consider including an item, say item i, in a different
test, we would have another true score ζ', monotonously related to
the true score of the present test. The function $P_i(\zeta')$ would have
another form than the function $P_i(\zeta)$.

The true-score scales of the different tests can be considered functions of one underlying latent trait. Let us denote the latent trait
value by the symbol θ. Now we can write the probability of a correct
response to item i as $P_i(\theta)$. The function $P_i(\theta)$ does not depend on the
test form in which item i happens to be included. It is assumed that
the function $P_i(\theta)$ depends on a number of item parameters. Several
one-factor models for dichotomous items have been proposed, like the
Rasch model (Rasch, 1960), and the two-parameter and three-parameter
logistic models (Birnbaum, 1968). These models are examples of
unidimensional *item response models* (IRT models); there are also

multidimensional item response models and models for more than two response categories (some examples will be given in Chapter 9).

The probability of occurrence of a particular response pattern on a n-item test given the latent trait score θ can be written as the product

$$P(X_1 = x_1,...,X_n = x_n \mid \theta) = \prod_{i=1}^{n} P_i(X_i = x_i \mid \theta)$$

$$= \prod_{i=1}^{n} P_i(\theta)^{x_i}[1-P_i(\theta)]^{1-x_i}$$

(6.16)

where $x_i = 1$ for a correct response, and $x_i = 0$ for an incorrect response.

We can estimate the item parameters of the item characteristic curve (ICC), $P_i(\theta)$, of item i from responses to the test. Next, we can compute the true score for a given value of θ as:

$$\tau = \sum_{i=1}^{n} P_i(\theta)$$

(6.17)

The conditional error variance for a given true score can be computed as

$$\sigma^2_{E|\tau} = \sigma^2_{E|\theta} = \sum_{i=1}^{n} P_i(\theta)[1- P_i(\theta)]$$

(6.18)

where θ is the latent ability that corresponds with the true score.

Further, it is possible to estimate the population distribution of θ (Bock and Aitkin, 1981). When an estimate of the population distribution is available, we can compute a Bayesian point estimate of θ.

6.5 Item analysis and item selection

In traditional item analysis, the most common indexes that are computed are those for item difficulty and item discrimination power. We can do likewise for a nested design as well as for a crossed design. Here, we discuss the computation of item statistics within the context of a crossed design with N persons and n items.

For each item, we can compute the mean item score. For dichotomous items, the mean score is equal to the proportion correct, or item difficulty index p_i. The higher the value of the item difficulty index is, the easier the item. The variance for item i is

$$Np_i(1 - p_i)/(N - 1) \approx p_i(1 - p_i) \tag{6.19}$$

The extent to which the item discriminates between high-scoring persons and low-scoring persons, the item's discriminating power, is approximated by the item-test correlation r_{it}. With relatively large tests, total test score is close to the true score, and the item–test correlation gives a fair impression of the item discriminating power. With small tests we have a problem. The correlation between item and test, r_{it}, is *spurious*: The measurement errors of the item and the test are correlated because the item is part of the total test. In this situation, it is better to use r_{ir}, the correlation between the item and the rest score, the total score minus the item score. This coefficient can be written as

$$r_{ir} = \frac{s_t r_{it} - s_i}{\sqrt{s_t^2 - 2s_i s_t r_{it} + s_i^2}} \tag{6.20}$$

When in the computation of the variances the numerator is divided by N, the item-rest correlation r_{ir} of dichotomous items can be written as

$$r_{ir} = \frac{M_+^{(i)} - M^{(i)}}{\sqrt{s_t^2 - 2s_i s_t r_{it} + s_i^2}} \sqrt{\frac{p_i}{1 - p_i}} \tag{6.21}$$

where p_i = the item proportion correct or item difficulty of item i, $M^{(i)}$ = the average score on the test minus item i, and $M_+^{(i)}$ = the average score on the test minus item i for the subgroup with item i correct.

A coefficient corrected for spuriousness and attenuation was suggested by Henrysson (1963), with coefficient α as estimator of test reliability.

In a homogeneous test, the two item indexes, item difficulty and item–rest correlation, give us information on the quality of the item. If necessary, screening of items can be done using these two indexes, at least when the sample is large enough to give relatively accurate sample estimates of these indexes. In a heterogeneous test, a test from which several subtests can be constructed, the item–rest correlation is less informative. With heterogeneous tests consisting of several subtests, factor analysis methodology and possibly structural equation modeling, are approaches that might be useful for test construction and test development in general, and item analysis and item selection in particular. This, however, is beyond the scope of the present chapter (see, e.g., McDonald, 1999).

The item–rest correlation r_{ir} should have at least a positive value, the higher the values of the correlation, the better. An item with a value close to 0 may suppress reliability when included in the test, if an unweighted sum score is used. The advantage of unweighted scores is that they are simple, easy to defend, and not sensitive to sample fluctuations. Optimal weights might be obtained from an IRT analysis. Items with a low discriminating power might be rejected for selection in a final test version.

Although IRT models are discussed in Chapters 9 and 10, here some remarks will be made about some of the dichotomous IRT models in the context of item analysis and item selection.

In the Rasch model, all items are assumed to be equally discriminating. Item selection within the Rasch model involves selecting items with similar item discriminations. In item selection, relatively undiscriminating items are deleted from the test, because they do not fit the model. However, an item with a better than average discrimination will be rejected in a Rasch analysis. Is this desirable from a practical point of view?

What is the optimal difficulty level of test items? Is it good to have items that differ strongly in difficulty level or not? The answer to this question depends on the purpose of the test and the discriminating power of the items. Let us assume that the purpose is to discriminate well in a population of persons. Let us also assume that the items are strongly discriminating. Then the probability of a correct answer shows a large increase at a particular level of the latent trait. In Figure 6.2 we have two such items. The probability of a correct answer on item 1 is close to 1 for levels of the latent trait for which the probability of a correct answer on item 2 is still close to zero. These two items define a Guttman scale as long as no other items of intermediate difficulty are chosen for inclusion in the scale. In the perfect Guttman scale, the probability of a correct answer is zero or one: at a particular level of the latent trait the probability jumps from zero to one. That is to say, the Guttman model, leading to the perfect Guttman scale, is a pathological probability model or deterministic model for dichotomous item

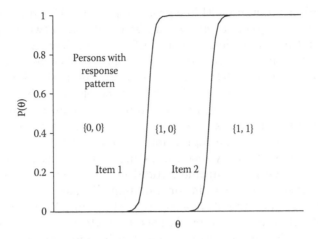

Figure 6.2 Two strongly discriminating items.

responses. The Guttman model can also be conceived of as a typical proto-IRT model.

For comparison with Figure 6.2, two less-discriminating items are displayed in Figure 6.3.

If we want to discriminate between persons within a broad range of θ, we better choose items of distinct difficulty levels when we have highly discriminating items like those in Figure 6.2. Each item then contributes

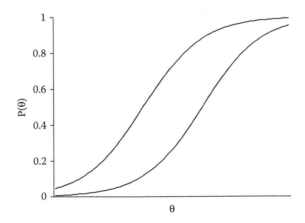

Figure 6.3 Two items with a moderate discriminating power.

to a finer discrimination within a group of persons. The group of persons with two items correct out of two can be divided into two subgroups by including a third item that is more difficult than item 2.

In practice, most items are more similar in discriminating power to the items in Figure 6.3 than to the items in Figure 6.2. An impression of the discriminating power of items can be obtained by plotting the item-test regression in a figure like Figure 6.1. In case all items would have been Guttman items, the item-test regression would have looked quite differently from the regression in Figure 6.1.

With more moderately discriminating items, it proves to be better to select items with comparable difficulties. If we want to discriminate between persons in a population an item difficulty of about 0.50 is optimal unless guessing plays a role (Cronbach and Warrington, 1952). A test with this kind of items is less accurate for persons with very high and very low latent trait values, but for most persons the test is more accurate than a test with spread item difficulties. Some item selection procedures, however, automatically select items with spread difficulties. In a procedure for scale construction proposed by Mokken (1971), the scale is not formed by deleting items that are not satisfactory, but by step-wise adding items that satisfy certain criteria. The procedure starts with the selection of the items most different in difficulty if the items do not differ with respect to discriminating power (see Mokken, Lewis, and Sijtsma, 1986); see Croon (Croon, 1991) for an alternative procedure. More information on procedures for test construction is presented in Exhibit 6.4.

Exhibit 6.4   Item selection in test construction: Some practical approaches

Traditional test construction relies heavily on the indexes for item difficulty and item discriminating power. In addition, item correlations can be taken into account in the construction of tests. Also, if some external criterion is available, item validity (i.e., the correlation of item and criterion scores) can be used.

Several methods have been proposed to construct a relatively homogeneous test from a pool of items. One possible classification of methods is the following:

1. Step-wise elimination of single items or subsets of items. Eliminate those items that do not correlate with the other items (e.g., set a certain threshold for an acceptable item—rest correlation). Repeat

the procedure after elimination of items until the desired standard is reached. The contribution of the item to test reliability can also serve as a criterion for elimination of an item.

2. Step-wise addition of single items or subsets of items. The construction of the scale starts with the two items that have the strongest relationship according to a particular index. Next, the item with the strongest relation to the items in the scale in formation is added if certain conditions are satisfied. The process is repeated until no further items are eligible for inclusion. The whole process can be repeated with the construction of the next scale from the remaining items. Another technique in this class of procedures is hierarchical cluster analysis, based on, for example, the average intercorrelation between clusters (see also Nandakumar, Yu, Li, and Stout, 1998). In hierarchical cluster analysis, scales are constructed simultaneously.

3. Item selection can be based on item correlation with an external criterion. The external criterion can be a classification in a diagnostic category (e.g., people with schizophrenia). Although this procedure produces a useful instrument for diagnostic purposes, it does not guarantee the construction of a homogeneous scale.

4. Factor analysis of item intercorrelations. Usually this approach is applied when several factors are thought to underlie item responses. Traditional factor analysis is sometimes difficult to apply with dichotomous items. An obvious way out is to use one of the procedures of nonlinear factor analysis (Panter, Kimberly, and Dahlstrom, 1997). Nonlinear factor analysis can be viewed as a multidimensional IRT analysis. IRT will be outlined in Chapters 9 and 10.

If guessing plays a role, we should take that into account. Let us have a test with multiple-choice items. An item has k response options and one of these is correct. Let us further assume that a person knows the answer to an item and responds correctly or does not know the answer and guesses randomly. The probability of a correct response under random guessing equals $c = 1/k$. Then the relation between p', the item of difficulty under guessing, and p, the item difficulty without guessing, equals

$$p' = c + (1 - c)p \qquad (6.22)$$

From this it follows that the optimal difficulty for a multiple-choice item with four options is about 0.625. Actually, the optimal value is likely to be somewhat higher (Birnbaum, 1968).

In case we are interested not so much in discriminating between persons as in comparing persons with a standard, the answer to the question of optimal item difficulty is a different one. Assuming that we are interested in a fine discrimination in the neighborhood of a true-score level equal to τ_0, an item has an optimal difficulty if at τ_0 the probability of a correct answer equals $c_i + (1 - c_i) \times 0.5$ for c_i equal to 0 and a bit higher if c_i is larger than 0.

This result is obtained with an IRT analysis (Birnbaum, 1968). Such an analysis is to be preferred to an analysis within the context of classical test theory. The true-score scale is defined in terms of the items that constitute the test. If one item is dropped from the test, the true score on the test changes as well as the value of τ_0. Later we will discuss test construction more fully in terms of IRT (see Chapter 10).

The outcome of an item analysis in classical test theory not only depends on the test that includes the item. The sample of persons who have answered the test determines the estimates of item discriminating power and item difficulty. It is important to remember this when evaluating test results from different groups of examinees. The groups might differ with respect to performance level, and, consequently, an item might have a different estimate of difficulty level in each group. Item selection and test construction on the basis of test statistics such as proportion correct is not justified when the estimates for different items come from incomparable groups.

Exercises

6.1 Compute the probability that a person with a domain score equal to 0.8 answers at least 8 out of 10 items correct, assuming that the items have been randomly selected from a large item pool.

6.2

    a. Compute the proportion correct and the item–rest correlation of item 8 in the table of Exercise 5.1. Compute the item-test regression of this item.

    b. Compute the item–rest correlations of the remaining items as well. Which item should be dropped first when a scale is constructed by a step-wise elimination of items?

6.3 In a testing procedure, each examinee responds to a different set of ten items, randomly selected from a large item pool. The test mean equals 7.5, and standard deviation equals 1.5. What might be concluded about the test reliability?

6.4 For a person p the probability of a correct answer to two items is $P_1(\zeta_p) = 0.7$ and $P_2(\zeta_p) = 0.8$, respectively. Compute the probabilities of all possible response patterns.

6.5 What information would you like to obtain in order to verify whether the assumptions made by Keats, see Equation 6.14 and Equation 6.15, are realistic?

6.6 A test consists of three items. The probabilities correct for person p are $P_1(\zeta_p) = 0.6$, $P_2(\zeta_p) = 0.7$, and $P_3(\zeta_p) = 0.8$. Compute the error variance on the total score scale. Also compute the error variance under the binomial model assumption. Comment on the difference.

6.7 Compare r_{it} and r_{ir} for tests with all item variances equal to 0.25 and all interitem covariances equal to 0.05. Compute the correlations for test lengths 10, 20, and 40.

CHAPTER 7

Validity and Validation of Tests

7.1 Introduction

In scientific inquiry, validity of statements refers to the degree to which there is empirical evidence to support the adequacy and appropriateness of these statements. More specifically, a measure is valid to the extent that it measures what is intended to measure. How vague this description may be, it makes clear that validity is not a property of a measurement instrument, but rather the interpretation of test scores and their use. In other words, how adequate and appropriate are the interpretations and uses of test scores, taking into account empirical evidence and, eventually, theoretical rationales. Validation, then, is the process through which the validity of the proposed interpretation of scores is investigated. The process of validation amounts to collecting empirical evidence to provide stable and generally accepted theoretically based interpretations of test scores (and other modes of assessment, for that matter).

In the present chapter we will first go into the problem of validity as a specific term—that is, validity of test scores or other assessments. Following the developments in the conceptualization of validity, validity as a unitary concept will be highlighted. Unified validity integrates earlier forms of validity of test scores by focusing on the various sources of evidence that might be used in evaluating a proposed interpretation of test scores for particular purposes. To date, Kane (2006) provides a relevant overview of validity and validation of tests and other measurement instruments in the social and behavioral sciences. In Exhibit 7.1 a sketch will be given with respect to validity and the *Standards*, in Section 7.2 the various sources of evidence are provided. How important these sources may be, in the present monograph it is impossible to deal with them extensively. We stick to the statistical aspects of validity and validation. In Section 7.3 selection effects in validation studies are outlined, in Sections 7.4 and 7.5 classification, and in Section 7.6 on what has been coined the evidence-based approach using analyses of the relationship of test scores to variables external to the test (*Standards*, APA et al., 1999, pp. 13–14). Some remarks will be made on validation and item response theory (IRT) in Section 7.7, while research validity, a form of validity not typical for test scores, appears in Section 7.8. An important topic that is not included is validity generalization. Suffice it to mention *The Handbook of Research Synthesis* (1994) edited by Cooper and Hedges, *Methods of Meta-Analysis* (1990) by Hunter and Schmidt, and Van den Noortgate and Onghena (2005), who include references to software in their overview of the field.

Exhibit 7.1   The many faces of validity and the standards

Where it all started is not easy to trace. Validity, it seems, has been and will be a perennial theme to scrutinize and discuss for test theorists and practitioners in the field of psychological and educational testing. Let us only mention a few highlights in the history of the conceptualization, operationalization, and periodic canonization of test validity.

In the 1950s, diverse forms of validity were proposed to fit different situations. The APA (1954) and the AERA (1955) mention four types of validity: content, predictive, concurrent, and construct. Although there was consensus, there were dissidents: Anastasi (1954) added face validity, factorial validity, and various types of empirical validity, Mosier (1947) analyzed face validity into validity by assumption, validity by definition, the appearance of validity, and validity by hypothesis. Interestingly, Ebel (1961) already stressed the evidence base of validity (more explicitly formulated decades later in the 1999 *Standards*). The seminal paper of the 1950s, with the benefit of hindsight, is L. J. Cronbach and P. E. Meehl's *Construct Validity of Psychological Tests* (1955).

In 1966 the APA published the *Standards for Educational and Psychological Tests and Manuals*, explicitly using the term *standard* (i.e., level or degree of quality that is considered proper or acceptable) (APA, 1966). Later editions of the *Standards* provide a frame of reference to assure that relevant issues are addressed. Also in the 1974 Standards (APA, AERA, and NCME 1974), the distinction in four types of validity mentioned above remains, now also endorsed by AERA and NCME.

A next step in the long march toward a unified view of validity is the 1985 *Standards* (APA, AERA, and NCME, 1985), greatly expanding the formalization of professional standards for test use. No longer are types of validity distinguished, but rather categories of validity evidence called content-related, criterion-related, and construct-related evidence of validity. Messick (1989, pp. 18–20) sketches the historical trends in conceptions of validity, and again, in 1994 made his last public plea for a unified view of validity (Messick, 1995), culminating in the 1999 *Standards* (APA et al., 1999).

The later codification of 1999 is still very useful today. But we must keep in mind that although a unified view of validity is surely a great stride in the long march, and although a listing of sources of validity evidence (see Section 7.2 and *Standards* 1999, pp. 11–17) is illuminating and useful, a validation study is always an empirical piece of research according to general rules of research methodology. Apart from the fact that the purpose of the *Standards* is to provide criteria for the evaluation of tests, testing practices, and the effects of test use, it is not a manual for how to set up a validation study with explicit consideration of specific designs and statistical analyses.

There are also researchers who are critical of the codification. Borsboom, Mellenbergh, and Van Heerden (2004), for example, complain that it focuses on epistemology, meaning, and correlation rather than on ontology, reference, and causality. Validation research should be directed at response processes and should be driven by a theory of these processes. Study of response processes has also been stressed by the 1999 Standards. One may question whether there are generally accepted theories of response processes around.

7.2 Validity and its sources of evidence

"Validity refers to the degree to which evidence and theory support the interpretation of test scores entailed by proposed uses of tests." This is the opening sentence of the chapter on validity in the latest edition of the *Standards* (APA et al., 1999, p. 9). It is no definition in the Aristotelian sense (i.e., *per genum proximum et differentiam specificam*). Neither is it an operational definition: it does not explicitly refer to the relevant operations to ensure validity. The *Standards* therefore proceed by stating: "The process of validation involves accumulating evidence to provide a sound scientific basis for the proposed score interpretations" (l.c., p. 9). Fortunately, this is an explicit statement: the unified view of validity entails that validity is evidence based, and the sources of evidence are:

- test content
- response processes
- internal structure

- relations to other variables
- information on the consequences of testing;
- the latter evidence has to do also with social policy and decision making.

The evidence based on test content can be obtained by analyzing the relationship between a test's content and the construct it is intended to measure. Response processes refer to the detailed nature of performance. It generally comes from analyses of individual responses (e.g., do test takers use performance or response strategies; are there deviant responses on certain items, etc.). The evidence based on internal structure comes from the analysis of the internal structure of a test (e.g., can the relationships among test items be accounted for by a single dimension of behavior?). In Chapter 3 we already met the analysis of the internal structure of test items in the context of internal consistency reliability. And the latter form of reliability is worked out (and liberalized, so to say, from the assumptions of classical test theory) in the broader framework of generalizability theory. G theory, therefore, bridges the gap between reliability and validity (cf. Cronbach et al., 1972). Performance assessment is generally thought to have the right content, but it needs further validation (Messick, 1994; Lane and Stone, 2006).

The largest category of evidence is evidence based on relations to other variables. This category of evidence analyzes the relationship of test scores to external variables (e.g., measures of the same or similar constructs, measures of related and different constructs, performance measures as criteria). Instead of using the old-fashioned label of concurrent validity (e.g., the concept of validity in the unified view refers to the way evidence can be obtained for validity). The category based on relations to other variables includes the following:

- Convergent and discriminant evidence
- Test-criterion relationships
- Validity generalization

The first subcategory of convergent and discriminant evidence has its early beginnings with Cronbach and Meehl (1955) and, most importantly, with Campbell and Fiske (1959). This subcategory of what was called construct-related validity is presented in Section 7.5. Test-criterion relationships studies what has been called criterion-related validity, and still earlier, predictive validity. Validity generalization is the evidence obtained by giving a summing-up of earlier findings with respect to similar research questions (e.g., of the findings of criterion-related correlation studies, with the same or comparable dependent and independent variables). Validity generalization is also known under the terms meta-analysis, research synthesis, or cumulation of studies. A new development that should be mentioned is the argument-based approach to validity. One could call this the hermeneutic or interpretative argument as Kane (2006, pp. 22–30) has it. This development is too fresh to include it in the present chapter.

So far, it is all rather abstract. How can it be made more concrete? How do we proceed in the validation of a test? Ironically, to make it clear how we study validity empirically, we do better to go back to the 1985 *Standards* trichotomy of test validity.

The following are the three validities in the 1985 *Standards*:

1. Content-related validity: In general, content-related evidence demonstrates the degree to which the sample of items, tasks, or questions on a test is representative of some defined universe or domain of content.
2. Criterion-related validity: Criterion-related evidence demonstrates that scores are systematically related to one or more outcome criteria. In this context, the criterion is the variable of primary interest, as is determined by a school system, the management of a firm, or clients, for example. The choice of the criterion and the measurement procedures used to obtain criterion scores are of central importance. Logically, the value of the criterion-related study depends on the relevance of the criterion measure that is used.
3. Construct-related validity: The evidence classed in the construct-related category focuses primarily on the test score as a measure of the characteristics of interest. Reasoning ability, spatial visualization, and reading comprehension are constructs, as are personality characteristics such as sociability and introversion. Such characteristics are referred to as constructs because they are theoretical constructions about the nature of human behavior (APA et al., 1985, pp. 9–11).

Each of these validities leads to methods for obtaining evidence for the specific type of validity. The methods for content-related validity, for example, often rely on expert judgments to assess the relationship between parts of the test and the defined universe. This line of thinking or approach is embedded in generalizability theory as discussed earlier. In addition, certain logical and empirical procedures can be used (see, e.g., Cronbach, 1971).

Methods for expressing the relationship between test scores and criterion measures vary. The general question is always: how accurate can criterion performance be predicted from test scores? Depending on the context, a given degree of accuracy is judged high or low, or useful or not useful. Two basic designs can be distinguished for obtaining information concerning the accuracy of test data. One is the *predictive study* where test data are compared (i.e., its relationships are studied with criterion scores obtained in the future). The second type of study is the so-called *concurrent study* in which test data and criterion data are obtained simultaneously.

The value or utility of a predictor test can also be judged in a decision theory framework. This will be exemplified in a later section. There, errors of classification will be considered as evidence for criterion-related validity.

Empirical evidence for the construct interpretation of a test may be obtained from a variety of sources. The most straightforward procedure would be to use the intercorrelations among items to support the assertion that a test measures primarily or substantially a single construct. Technically, quite a number of analytical procedures are available to do so (e.g., factor analysis, multidimensional scaling (MDS), IRT models). Another procedure would be to study substantial relationships of a test with other measures that are purportedly of the same construct, and the weaknesses of the relationships to measures that are purportedly of different constructs. These relationships support both the identification of constructs and the distinctions among them. This quite abstract formulation is taken from the *Standards* (APA et al., 1985, p. 10). In a later section the so-called multitrait–multimethod approach to construct validation will be considered more concretely and in more detail.

Before going into certain aspects and procedures for validation studies, it is important to consider the problem of selection and its effects on the correlation between, for example, test X and criterion Y—that is, the (predictive) validity of test X with respect to criterion Y. Essentially, this is applying statistics in the field of psychometrics: What is the influence of restriction of range on the value of the validity of a test?

7.3 Selection effects in validation studies

Suppose we want to study the validity of test X with respect to criterion Y. We already use the test for selection. Only persons with a score x larger than or equal to score X_c on the test are admitted or selected, and we have criterion scores Y only for these selected persons. We are interested in the correlation between X and Y within the total population, but we can compute the correlation only for the subpopulation of selected persons. Is it possible to estimate the correlation within the total population?

We will derive the relation between the correlation in the subpopulation and the correlation in the total population following a practice suggested by Gulliksen (1950). Lower-case characters designate statistics in the subpopulation, and capitals designate statistics in the total population. We assume that the regression of Y on X is linear and that the regressions are identical in the total population and the subpopulation. In addition, we assume that the variances of estimation errors (i.e., the variances around the regression line of Y on X) are identical. These two assumptions can be expressed mathematically as

$$\frac{S_Y}{S_X} R_{XY} = \frac{s_y}{s_x} r_{xy} \tag{7.1}$$

and

$$S_Y^2 \left(1 - R_{XY}^2\right) = s_y^2 \left(1 - r_{xy}^2\right) \tag{7.2}$$

Now we have two equations with two unknowns, the criterion variance in the total population S_Y^2 and the correlation between predictor X and criterion Y in the total population.

We can solve Equation 7.1 for S_Y^2 and substitute the result in Equation 7.2. Next, we can solve this equation for R_{XY}. The result is

$$R_{XY}^2 = \frac{r_{xy}^2}{r_{xy}^2 + \frac{s_x^2}{S_X^2}\left(1 - r_{xy}^2\right)} \tag{7.3}$$

Selection on test X not only depresses the correlation of this test with the criterion, but due to incidental selection on other variables, the correlations of other variables are affected as well. The pattern of correlations differs between the subpopulation and the total population; see Gulliksen (1950) or Sackett and Yang (2000) for the case of more than two variables in selection. The lowering of the correlation of the explicit selection variable X sets this test at a disadvantage when it is compared with a competing test in the selected group.

The value of the correlation between two measurement instruments in a subpopulation can be smaller than the correlation in the total population, even though no explicit selection has taken place. Self-selection of persons has a similar effect as selection to the extent to which two variables are correlated. Let us give an example of a situation where both selection and self-selection might operate. Assume that mathematical ability is an important ability for a particular study. It is reasonable to assume that there is a relationship between mathematical ability and achievement in the study. We correlate achievement with an ability measure only to find a low correlation. The low value does not invalidate the hypothesis of a relationship. The low correlation might be due to the combined effects of selection and self-selection.

7.4 Validity and classification

The size of the correlation between a predictor and a criterion sometimes says very little about the utility of the predictor (Taylor and Russell, 1939). We will discuss this using Figure 7.1.

Assume that we want to hire a fixed percentage of applicants on the basis of their scores on a predictor X. For a high score on the predictor we accept the applicant, for a low score we reject the applicant. We have a criterion Y with the categories *satisfactory* and *unsatisfactory*.

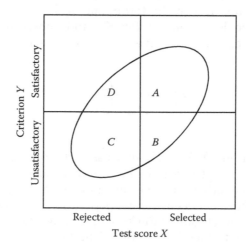

Figure 7.1 Classification into satisfactory/unsatisfactory and rejected/accepted.

We have four different outcomes of the selection procedure with corresponding proportions:

- The proportion A is accepted and satisfactory.
- The proportion B is accepted but is unsatisfactory.
- The proportion C is correctly rejected.
- The proportion D is rejected although these applicants are satisfactory.

For the sake of simplicity, we will not bother about the way in which we can make the distinction between the two groups of rejected applicants. The proportion $A + B$ is called the *selection ratio*; this proportion was assumed to be fixed in the example. The proportion $A + D$ is known as the *base rate*; naturally, the base rate is fixed. The proportion $(A + C)$ is called *classification accuracy*.

Taylor and Russell assumed that a bivariate normal distribution underlies the double dichotomy *satisfactory/unsatisfactory* and *rejected/accepted*. The correlation between the continuous variables X and Y is denoted by r_{XY}. Given a base rate equal to 0.50, a selection ratio equal to 0.30, and a validity coefficient r_{XY} of the underlying continuous variables equal to 0.50, a *success ratio* $R = A/(A + B)$ equal to 0.74 is obtained. We might compare this outcome with the expected outcome if we had not used predictor X. If we had selected the persons randomly, we would have obtained a success ratio equal to the base rate (0.50). We might compare the success ratio with the success ratio that we would have obtained with a perfect predictor. In the present case, the success ratio with a perfect predictor would have been equal to 1.00. The utility of the test as a selection instrument is higher than we might have expected from the size of the validity coefficient.

The results obtained by Taylor and Russell depended very much on their choice of the success ratio as a measure of test efficiency and the fact that the selection ratio was fixed. We have two kinds of errors: incorrectly accepting persons and incorrectly rejecting persons. In the Taylor and Russell approach, only one kind of classification error counted: incorrectly accepting applicants. In many other applications, both kinds of classification errors should be considered and then the utility of the test may be quite different. Let us from now on assume that both kinds of classification errors are relevant and that the selection ratio is not fixed.

Taylor and Russell divided persons from a single population into two groups on the basis of criterion performance. Sometimes we are

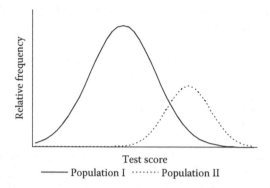

Relative frequency

Test score

——— Population I ········ Population II

Figure 7.2 The distribution of test scores within two populations.

dealing with two or more different populations and we have to decide to which population a person belongs. For this purpose, we might use a predictor X. In its most simple form the problem comes down to deciding on a cut score on the predictor. Persons with a score equal to the cut score or higher are classified as belonging to one of the populations, the others are classified as belonging to the other population. In Figure 7.2 the classification problem is illustrated for the case of two populations.

Let us assume that we have two populations: population I and population II. The purpose of the test is to detect population-II persons or type-II persons, as they need treatment. Let us assume that we have a cut score X_c. All persons with a score equal to the cut score or higher than the cut score are classified as type-II persons. We make classification errors. Some persons are incorrectly classified as type-I persons, others are incorrectly classified as type-II persons. With a low cut score we have a relatively large number of false positives, persons who do not belong to population II, but are incorrectly classified as type-II persons. The term *positive* originates from medicine where it indicates the presence of a condition (illness) for which it is screened. In our example *positive* is associated with relatively high test scores. For a high cut score we have, in contrast, a relatively large number of false negatives, persons for whom the diagnosis "type-II" has been missed. At the score X at which the curves for the two distributions meet, the two kinds of errors are in balance. We might choose this value of X as X_c.

The point X at which the probability of belonging to population II given the observed score, the posterior probability $P(\text{II} \mid x)$, equals the

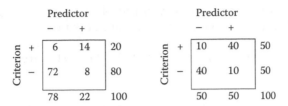

Figure 7.3 Classification tables for two base rates.

posterior probability of belonging to population I depends on the base rate $P(\text{II})$. The way base rate influences the posterior probabilities is easily seen by applying Bayes' theorem:

$$P(\text{II} \mid X) = \frac{P(X \mid \text{II})P(\text{II})}{P(X)} = \frac{P(X \mid \text{II})P(\text{II})}{P(X \mid \text{I})P(\text{I}) + P(X \mid \text{II})P(\text{II})} \qquad (7.4)$$

One might imagine that it is difficult to detect type-II persons when the base rate is low. The accuracy of the measurement instrument with respect to the detection of type-II persons is called its sensitivity. Sensitivity is defined as the proportion of persons with a disorder for whom the diagnosis is correctly made. Not only the sensitivity of the measurement instrument is important, but also its specificity. The specificity of a measurement instrument is defined as the proportion of healthy persons for which the diagnosis is correctly rejected. The definitions of the two concepts can be illustrated with the data in Figure 7.3. On the left-hand side of this figure the base rate is 0.20. The diagnosis (+) is correctly made in 14 of the 20 cases. So, the sensitivity of the procedure is 0.70. The diagnosis is correctly rejected in 72 of the 80 cases. So, the specificity of the procedure is 0.90. On the right-hand side of the figure the base rate is 0.50. The sensitivity is 0.80. The specificity is also 0.80.

What should we do when missing the diagnosis "type-II" is considered to be worse than the incorrect classification of type-I persons? When the difference between the losses associated with the classification errors is small, the same cut score might be optimal: the cut score might be robust against a small change in the losses. For a larger difference the cut score should be adapted.

Missing a type-II person might be twice as serious as misclassifying a type-I person. In that case, we are ready to categorize a subject as a type-II person as long as twice the posterior probability of type-II exceeds the posterior probability of type-I—that is, as long as $P(\text{II} \mid X)$ exceeds one third. The general classification rule is

$$lP(\text{II} \mid X) > P(\text{I} \mid X) \text{ or } P(\text{II} \mid X) > 1/(l + 1) \Rightarrow \text{classify as type-II}$$
$$(7.5a)$$

$$lP(\text{II} \mid X) < P(\text{I} \mid X) \Rightarrow \text{classify as type-I} \qquad (7.5b)$$

where l is the ratio of the loss associated with misclassifying a type-II person and the loss associated with misclassifying a type-I person.

When classification errors are serious or when many classification errors are made, one might decide to use test X only as the first screening device. In that case, two cut scores might be used. If a person scores high or low a final classification is made. Scores in between fall in the category *yet undecided* (Cronbach and Gleser, 1965). For other and more difficult classification problems see Hand (1997).

Let us now return to the original example with the classification accepted/rejected and satisfactory/unsatisfactory. Instead of two populations there is only one population. Persons with a criterion score equal to or larger than Y_c are considered to be satisfactory; persons with lower scores are considered unsatisfactory. The utility of the test depends on the cut score between *accept* and *reject*, X_c. With a high cut score more persons are incorrectly rejected; with a low cut score more persons are incorrectly accepted. The situation is comparable to that discussed in connection with two populations. In this case decisions must be made on the basis of posterior distributions $P(Y \mid X)$.

Which cut score is optimal depends on the seriousness of the classification errors. To simplify matters, we might consider a discrete loss. Accepting an applicant who is unsatisfactory is equally serious for all those applicants. Rejecting applicants is equally serious for all those applicants incorrectly rejected. One possibility is that incorrectly accepting an applicant is equally serious as incorrectly rejecting an applicant. But it is also possible that one kind of classification error is considered to be more serious than the other kind of classification error. We should first determine the losses associated with the two kinds of errors or rather the ratio of these two losses. When we have the loss ratio it is possible to obtain the optimal cut score for a given bivariate distribution of test and criterion scores. For bivariate normally distributed test and criterion scores the optimal cut score is given by Alf and Dorfman (1967). If the two kinds of errors are equally serious, the optimal cut score is the value X for which the expected

Y equals Y_c, the value of Y on the border between *satisfactory* and *unsatisfactory*. This is easily verified. For this score X the proportion of satisfactory Y equals the proportion of unsatisfactory Y due to the normality of the distribution of Y given score X.

It might seem too simple to regard all decisions where persons are incorrectly accepted as equally serious. It might seem more adequate if the seriousness of the classification error depends on the value on the criterion. Van der Linden and Mellenbergh (1977) introduced a linear loss function with decisions on passing and failing examinees. Cronbach and Gleser (1965) gave a systematical treatment of decision making using tests. Petersen and Novick (1976) discussed decisions in the context of culture-fair selection.

In criterion-referenced measurement (Hambleton and Novick, 1973; Popham and Husek, 1969), we are interested in the domain score of an examinee. It is assumed that the examinee has mastered the subject matter if the domain score is at least as high as a standard set on the domain score scale. A test X is used to verify whether an examinee has mastered the subject matter. When the test score is high, the examinee passes the test; when the test score is low, the examinee fails the test. Now, we may look again at Figure 7.1. Instead of criterion Y we have the domain score. Instead of the dichotomies *satisfactory/unsatisfactory* and *rejected/accepted* we have the dichotomies *mastery/nonmastery* and *pass/fail*.

In criterion-referenced measurement, it is possible to have a homogeneous population of examinees with a high mastery level. For a reasonable test length, reliability is low with such a population. There is a strong regression effect: The expected true score given a low observed score is strongly shifted toward the mean.

As a consequence of the strong regression effect, the optimal cut score for "pass" might be relatively low when the average performance on the test exceeds the standard (for standard setting, see Exhibit 7.2). Then the question arises whether we should use the low "optimal" cut score and risk a negative effect on the study commitment of new groups of examinees. In criterion-referenced measurement, the problem of low test reliability due to the homogeneity of the population has led to suggestions of alternative coefficients. For example, coefficient kappa (Cohen, 1960; for standard errors: Fleiss, Cohen, and Everitt, 1969) has been suggested as an index for decision consistency (Subkoviak, 1984). It should be noted, however, that decision consistency also might be low when the average mastery level of a homogeneous population is close to the standard of performance.

Exhibit 7.2   Standard setting for performance

In educational assessment, the uses and interpretations of standards are essential; hence, attention has to be paid to standard setting methods for test performance. Performance standards may be defined as the scores that must be achieved by examinees to be classified as, say, proficient. Consequently, a critical step in the use of an assessment procedure is to establish cut points dividing the score range into categories that are meaningful for the educational community (cf. *Standards*, APA et al., 1999, p. 53). An example is how to classify students on the basis of their score on National Assessment of Educational Progress (NAEP) Math Assessments into categories (a) basic, (b) proficient, and (c) advanced. A plethora of standard-setting methods have been developed. Different methods lead to different results. It is not likely that repeated application of the same method in connection with different tests gives equivalent results. Standard setting is an important and inevitable activity, but also an activity that remains based on the best subjective judgment of experts. A recent survey of standard-setting methods is presented by Hambleton and Pitoniak (2006).

When a new test version is introduced, a standard for this test version must be set. There are several possibilities to set the standard on a new test form, for example, using the standard-setting methods again. When the old test and the new test have items in common, one of the procedures for test equating (Chapter 11) might be applied.

The computation of coefficient kappa can be illustrated with Figure 7.4. In this figure the crosstabulation is given of the outcomes of two tests. Decision consistency, the proportion of identical decisions, is the sum of the proportions p_{11} and p_{00}. In coefficient kappa this proportion

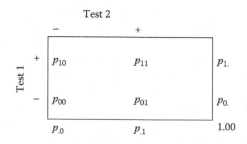

Figure 7.4 Decisions on two parallel tests.

is corrected for chance agreement. The coefficient for the two-by-two table is defined as

$$\kappa = \frac{(p_{11} + p_{00}) - (p_{.1}p_{.1} + p_{.0}p_{.0})}{1 - (p_{.1}p_{.1} + p_{.0}p_{.0})} \quad (7.6)$$

The decision consistency can be computed with the parallel-test method. In most applications, however, taking a parallel measurement is not practical or possible, and the proportions p_{11} and p_{00} must be estimated from the single administration of a test. The proportions $p_{.1}$ and $p_{.0}$ are set equal to the marginal proportions of the test. Theoretically, the best procedure for the estimation of the proportions p_{11} and p_{00} is the following (Huynh, 1978):

1. Estimate the distribution of domain scores $f(\zeta)$.
2. Estimate the conditional error distribution.
3. Compute the probabilities $p_{1|\zeta}$ and $p_{0|\zeta}$.
4. Compute the proportions $p_{11|\zeta} = p_{1|\zeta} \times p_{1|\zeta}$ and $p_{00|\zeta} = p_{0|\zeta} \times p_{0|\zeta}$.
5. Compute $p_1 = p_{.1} = \int p_{1|\zeta} f(\zeta) d\zeta$, $p_0 = p_{.0} = 1 - p_{.1}$, $p_{11} = \int p_{11|\zeta} f(\zeta) d\zeta$ and $p_{00} = \int p_{00|\zeta} f(\zeta) d\zeta$.
6. Compute κ.

Alternatively, the computations might be based on latent ability θ (Lee, Hanson, and Brennan, 2002).

7.5 Selection and classification with more than one predictor

When the quality of the classification or diagnosis based on a single measurement X is too low, other measurement instruments should be considered for inclusion in a test battery used for selection and classification purposes. What characteristics should potential additional tests have? In general, it seems wise to select tests that give new information useful for making the classification or diagnosis. Added tests should correlate with the criterion and have a low correlation with other predictors.

We will first demonstrate the point using a classical procedure of an unweighted sum of predictor scores. We will use the argument that has been used by Gulliksen (1950) with respect to the selection of items for a test. The validity of the sum of n scores X_i, X, with respect to criterion Y can be written as

$$r_{XY} = \frac{\displaystyle\sum_{i=1}^{n} s_Y s_i r_{iY}}{s_Y \displaystyle\sum_{i=1}^{n} s_i r_{iX}} = \frac{\text{ave}(s_i r_{iY})}{\text{ave}(s_i r_{iX})} \qquad (7.7)$$

where s_i is the standard deviation of the scores on test X_i. Let us assume that the variances of the tests do not differ much. In that case, a test that highly correlates with the criterion and not so much with the other predictors adds to the numerator and not to the denominator. When we select items for a test in order to maximize validity, items are selected that may decrease the reliability of the test.

The unweighted sum of predictor scores does not give the optimal combination of measurements. The obvious method is to use a weighted combination of scores. Optimal weights can be obtained from a multiple-regression analysis, where optimality is operationalized in terms of a least-square loss function. With scores on two predictors X_1 and X_2, and a criterion Y, the formula for the regression has the following form:

$$\hat{y}_p = a_0 + a_1 x_{1p} + a_2 x_{2p} \qquad (7.8)$$

The linear regression approach exemplifies the so-called compensatory model for selection. In this model the minimum requirement on the criterion is achieved by an additive combination of abilities. A low level of achievement for one ability can be compensated for by a high level for another ability. The compensatory model for selection with two predictors is displayed in Figure 7.5a. With errorless variables X_1 and X_2, the straight line gives all combinations of x_1 and x_2 that result in the critical criterion level. So, the straight line is the border between combinations of abilities x_1 and x_2 that correspond to a satisfactory criterion level (+), and combinations x_1 and x_2 that correspond to an unsatisfactory criterion level (–). The linear regression formula is adequate for the compensatory model of Figure 7.5a. The linear regression formula can be adapted to a degree of nonlinearity in the relation between the criterion and the predictors: powers of the

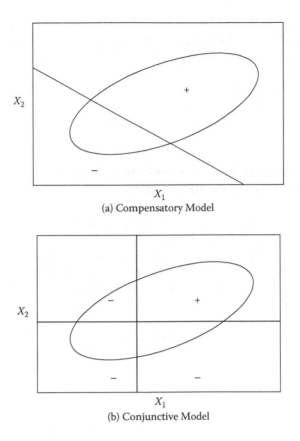

(a) Compensatory Model

(b) Conjunctive Model

Figure 7.5 Classification with two predictors.

predictor scores (x_1^2, etc.) can be added as predictors in the multiple regression formula.

Two other models may be discussed in the context of selection: the conjunctive model and the disjunctive model (Coombs, 1964). The conjunctive model requires persons to satisfy a minimum level of achievement on each of the relevant abilities. There is no possibility of compensation. The conjunctive model with two abilities X_1 and X_2 is illustrated in Figure 7.5b. The conjunctive model seems to ask for multiple cut scores as in the figure. Actually, the classification of examinees in the conjunctive model is more complicated than that when the predictors have reliabilities less than one. Then the score on one measurement instrument contains information with respect to the achievement on another measurement instrument, assuming that the measurement instruments are correlated. Lord (1962) demonstrated that the prediction of criterion performance is increased when some amount of compensation between the fallible measurements is allowed.

In the third model, the disjunctive model, persons may satisfy the criterion by having at least one sufficient ability. With two abilities X_1 and X_2, only the quadrant with both low X_1 and low X_2 is associated with unsatisfactory criterion performance.

7.6 Convergent and discriminant validation: A strategy for evidence-based validity

Evidence for construct-related validity can be obtained in quite a number of ways (i.e., using several research designs and corresponding statistical methods for data analysis). These methods in construct validation, proposed already by Cronbach and Meehl (1955) and more specifically by Campbell and Fiske (1959), may be summarized as follows:

- The study of group differences: If we expect two or more groups to differ on the test purportedly measuring a construct, this expectation may be tested directly resulting in evidence for construct-related validity.
- The study of correlations between tests and factor analysis: If two or more tests are presumed to measure some construct, then a factor analysis of the correlation matrix must reveal one underlying factor as an indicator of the common construct.
- Studies of internal structure: For many constructs, evidence of homogeneity within the test is relevant in judging validity.
- Studies of change over occasions: The stability of test scores (i.e., retest reliability) may be relevant to construct validation.
- Studies of process: Observing a person's process of performance is one of the best ways of determining what accounts for variability on a test (see, e.g., Cronbach and Meehl, 1955; Cronbach, 1990). In addition, judgment and logical analysis are recommended in interpretations employing constructs (cf. Cronbach, 1971, p. 475).
- An important elaboration and extension of the study of correlations between tests is Campbell and Fiske's convergent and discriminant validation by the multitrait–multimethod matrix.

7.6.1   The multitrait–multimethod approach

Suppose we have a few different measurement instruments of a trait. We might expect that these measurements correlate. If there really is an underlying trait, the correlations should not be too low. On the other hand, correlations between measurement instruments for different traits should not be too high; otherwise it makes no sense to make a distinction between the traits. In many investigations, several traits are measured using the same kind of instrument, for example, a questionnaire. This might pose a problem. When there is a correlation between two traits measured with the same method, we might wonder to which extent the correlation is due to the covariation of the traits and to which extent it is due to the use of a single measurement method. Campbell and Fiske (1959) proposed to use the multitrait–multimethod matrix research design in order to study the convergence of trait indicators and the discriminability of traits in validation studies. A hypothetical example with three traits and three methods is given in Figure 7.6.

The main diagonal contains the reliabilities. We might call these entries monotrait–monomethod correlations. In the first diagonal

		Method a			Method b			Method c		
		Trait 1	Trait 2	Trait 3	Trait 1	Trait 2	Trait 3	Trait 1	Trait 2	Trait 3
Method a	Trait 1	$r_{11}(aa)$	$r_{12}(aa)$	$r_{13}(aa)$	$r_{11}(ab)$	$r_{12}(ab)$	$r_{13}(ab)$	$r_{11}(ac)$	$r_{12}(ac)$	$r_{13}(ac)$
	Trait 2		$r_{22}(aa)$	$r_{23}(aa)$	$r_{21}(ab)$	$r_{22}(ab)$	$r_{23}(ab)$	$r_{21}(ac)$	$r_{22}(ac)$	$r_{23}(ac)$
	Trait 3			$r_{33}(aa)$	$r_{31}(ab)$	$r_{32}(ab)$	$r_{33}(ab)$	$r_{31}(ac)$	$r_{32}(ac)$	$r_{33}(ac)$
Method b	Trait 1				$r_{11}(bb)$	$r_{12}(bb)$	$r_{13}(bb)$	$r_{11}(bc)$	$r_{12}(bc)$	$r_{13}(bc)$
	Trait 2					$r_{22}(bb)$	$r_{23}(bb)$	$r_{21}(bc)$	$r_{22}(bc)$	$r_{23}(bc)$
	Trait 3						$r_{33}(bb)$	$r_{31}(bc)$	$r_{32}(bc)$	$r_{33}(bc)$
Method c	Trait 1							$r_{11}(cc)$	$r_{12}(cc)$	$r_{13}(cc)$
	Trait 2								$r_{22}(cc)$	$r_{23}(cc)$
	Trait 3									$r_{33}(cc)$

Figure 7.6 The multitrait–multimethod correlation matrix with three methods and three traits.

entry, for example, we have $r_{11}(aa)$, the reliability of the measurement instrument which measures trait 1 by means of method a. Adjacent to the main diagonal we have triangles with heterotrait–monomethod correlations. We also have blocks with correlations involving two different methods. Within these blocks, we have diagonals with correlations involving one trait. These monotrait–heteromethod values are the so-called validity diagonals; a gray background in the figure indicates the monotrait–heteromethod entrees.

According to Campbell and Fiske, a validation process is satisfactory if the following take place:

1. Correlations between measurements of the same trait with different methods are significantly larger than 0. Then we have convergence.
2. Correlations between measurements of a trait with different methods are higher than the correlations of different traits measured with the same method. The validity diagonals should be higher than the correlations in the monomethod–heterotrait triangles. In that case, we have discriminant validity.
3. A validity coefficient $r_{ii}(ab)$ is larger than the correlations $r_{ij}(ab)$ and $r_{ji}(ab)$.
4. In the heterotrait triangles of the monomethod blocks and the heteromethod blocks, the pattern of correlations is the same.

Campbell and Fiske considered only informal analysis and eye-balling techniques for the study of multitrait–multimethod matrices. Such matrices, however, may also be analyzed with generalizability theory (Cronbach et al., 1972). An alternative approach is to use confirmatory factor analysis. It belongs to the class of structural equation modeling (SEM), and is, among others, a promising procedure to obtain evidence of construct-related validation where more constructs are involved in a nomological network. Also, with more than one measure, confirmatory factor analysis with so-called structured means can be used to test hypotheses with respect to the tenability of equivalence conditions (e.g., strictly parallel measures, tau-equivalent measures) of a set of measures. Last but not least, this type of confirmatory factor analysis offers a fruitful approach to test validation. Technical details of the analysis of the multitrait–multimethod matrix by confirmatory factor analysis can be found in Kenny and Kashy (1992), while general details on alternative approaches can be found in Schmitt, Coyle, and Saari

(1977) and Schmitt and Stults (1986). The reader should, however, be warned: routine applications of SEM for the analysis of multitrait–multimethod matrices are doomed to fail due to all the pitfalls in the use of SEM. The lesson from all this is that none of the analytic approaches to multitrait–multimethod matrices should be done routinely. A thoughtful and well-balanced review of approaches to the multitrait–multimethod matrix has been given by Crano (2000).

Test manuals should provide information on reliability, validity, and test norms. The manuals cannot be exhaustive, however. After publication of a test, new research adds to the validation of test uses. Summaries of research and critical discussions of tests are needed. The Mental Measurement Yearbooks fulfill such a function. Let us take the Beck Depression Inventory (BDI), a frequently cited inventory. The BDI is reviewed by two reviewers in the *Thirteenth Mental Measurement Yearbook*, Carlson (1998) and Waller (1998). The BDI is a brief self-report inventory of depression symptoms. It has 21 items, scored on a four-point scale. The test is used for psychiatric patients. It also is frequently used as a screening device in healthy populations. The manual gives information on reliability, validity, and test norms. But, the reviewers argue that the manual is too short. Much useful information must be found in other published sources. Several aspects of validity are discussed by the reviewers. The inventory has face validity; the items are transparent. The high face validity makes the inventory vulnerable to faking. Correlations with other tests have been computed and a factor analysis has been done. The inventory discriminates patients from healthy persons. Waller notes that the information with respect to discrimination validity is lacking. What is, for example, the correlation of the BDI with an anxiety measure, a measure of a different construct?

7.7 Validation and IRT

Item response theory (IRT) provides models in which the responses of subjects on the individual test items are modeled. IRT models not only allow for the estimation of person and item parameters, but are also a statistical test for how good the model fits the data. So when a unidimensional IRT model is assumed, the test of model fit informs us about the existence of a single construct or latent trait underlying the observed item responses. IRT models are discussed in later chapters, so here it suffices to state in general terms the nature of the construct-related validation using IRT. To date, this is a promising terrain of research (more can be found in Embretson and Prenovost, 1999, and the references mentioned there).

Considering construct-related validity in the context of IRT does not exhaust the validity issue in psychometrics at large. In the next section, research validity will be discussed in the broad context of empirical behavioral research.

7.8   Research validity: Validity in empirical behavioral research

Empirical behavioral research is a broader context of research than test research and development (R&D). Therefore, more general aspects of validation are involved, and the type of validity in this broader context has been coined *research validity* (see, e.g., Shadish, Cook, and Campbell, 2002).

The tenability of theories and the generalizability of findings from empirical research are influenced by four classes of validation:

1. Statistical conclusion validity: This is defined as the extent to which the design of the study is sufficiently sensitive or powerful to detect outcome effects. It addresses the question whether the relationship(s) observed in the sample are due to chance.

2. Internal validity: This is the extent to which detected outcome effects, viz. test scores, are due to the operationalized cause rather than to other rivaling causes. The question of internal validity is often rephrased as: Are there no alternative explanations for the detected effects in terms of, for example, changes in test scores?

3. External validity: This is defined as the extent to which the detected outcome effects (test scores for that matter) can be generalized to theoretical constructs, subjects, occasions, and situations other than those specified in the original study. In most instances, this type of validity is used to refer to the question of whether the test scores or other effect measures that were found in the sample can be assumed to exist in the population (or, a certain, well-defined population) as well.

4. Construct validity: This type of validity is defined similarly as in the trichotomy above, as the extent to which the theoretical constructs in a study have been successfully operationalized. In other words, does the measurement on a certain variable represent the phenomenon or propensity it is supposed to measure?

These and similar definitions of research validities can be found in textbooks on research methodology. Elaborations have been proposed; a succinct overview is given by Cook and Shadish (1994; see also Cook, Campbell, and Peracchio, 1990).

Although research validities are generally not to be considered as part of test theory per se, it is important and relevant to point out that each of the above four types of research validity may be under threat of one sort or another. Among these threats are history, maturation, testing, instrumentation, statistical regression toward the mean, and mortality. One approach to circumvent one or more of these threats is to choose the appropriate design of the study and proceed along the road of performing a generalizability study. So here we see that the demarcation of psychometric reliability and validity is blurred. Earlier, generalizability studies were treated as a liberalization of classical test reliability. It also serves as a vehicle for validation studies.

Exercises

7.1 In a study, test X is administrated to all persons. Test Y is administrated to a selection of persons. Within each group with the same score on X persons are randomly chosen for selection into the group that is administered test Y. The correlation between X and Y equals 0.8. The variance of the scores on X within the selection equals 36.0. The variance on X in the total group equals 16.0. Estimate the correlation between X and Y in the total group.

7.2 Given is a ten-item test with the following frequency distribution in two groups A and B:

Score	0	1	2	3	4	5	6	7	8	9	10
f_A	0.043	0.109	0.130	0.174	0.217	0.174	0.087	0.043	0.022	0.0	0.0
f_B	0.0	0.0	0.0	0.0	0.045	0.091	0.136	0.182	0.227	0.182	0.136

We want to use the test in order to classify persons in the future. Both kinds of errors are equally serious. At what test score should we take the decision to classify a person as a "B person" assuming that population A has four times the size of population B? Can you comment on your result?

7.3 What happens if the base rate of belonging to group B in Exercise 7.2 is 0.5 instead of 0.2?

7.4 We have a test with a mean equal to 75.0, a standard deviation equal to 8.0, and a reliability equal to 0.25. With the test we want to decide which examinees are masters and who are nonmasters. The criterion of mastery is 70.0 on the true-score scale. The errors of classifying masters and nonmasters incorrectly are equally serious. Compute the optimal cut score under the assumption that the observed scores and true scores have a bivariate normal distribution.

7.5 Compute coefficient κ for the data in the following table:

+	10	60
–	20	10
	–	+

Principal Component Analysis, Factor Analysis, and Structural Equation Modeling: A Very Brief Introduction

8.1 Introduction

More than one century of factor analysis, an approach to data analysis quite relevant for test theory, has resulted in a tremendously large volume of publications. In this chapter only a tip of the veil is lifted. Nevertheless, it is hoped what is unveiled serves its purpose in statistical test theory, and also whets the reader's appetite.

As principal component analysis is a basic technique similar to factor analysis proper, more attention will be paid to this type of analysis in Section 8.2. In Section 8.3 exploratory factor analysis is introduced. Section 8.4 discusses confirmatory factor analysis and structural equation modeling.

8.2 Principal component analysis (PCA)

Essentially, principal component analysis (PCA) is a data-reduction technique based on mathematical operations on relations between (many) variables in order to get more insight into the data. Data reduction means that a new representation of the relationship between variables is sought that is more parsimonious than the initial one, but without loss of relevant information.

What are principal components? This is illustrated in the simple case where we have two variables. In Figure 8.1 the relationship between two standardized variables X_1 and X_2 with a correlation $r(X_1,X_2)$ equal to 0.61 is presented by an ellipse. The first axis $Z_1 = \sqrt{1/2}X_1 + \sqrt{1/2}X_2$ represents the first principal component with a variance equal to 1.61 ($1/2\mathrm{var}(X_1) + \mathrm{cov}(X_1,X_2) + 1/2\mathrm{var}(X_2)$), the second axis

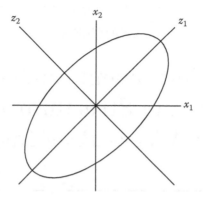

Figure 8.1 The ellipse representing the data for variables X_1 and X_2 with $r = 0.61$.

$Z_2 = -\sqrt{1/2}X_1 + \sqrt{1/2}X_2$ represents the second principal component, orthogonal to the first, with a variance equal to 0.39. The variances of Z_1 and Z_2 are referred to as eigenvalues and are denoted λ_1 and λ_2. The sum of the eigenvalues is equal to 2.0, the sum of the variances of the original variables X_1 and X_2.

The subject scores on X_1 and X_2 can be written as

$$X_1 = a_{11}F_1 + a_{12}F_2$$
$$X_2 = a_{21}F_1 + a_{22}F_2$$

where F_1 and F_2 are the principal components after standardization, and $a_{11} = 0.897$, $a_{12} = -0.442$, $a_{21} = 0.897$, and $a_{22} = 0.442$. Notice that

$$a_{11}^2 + a_{21}^2 = \lambda_1 = 1.61$$

and

$$a_{12}^2 + a_{22}^2 = \lambda_2 = 0.39.$$

The correlation matrix is perfectly reproduced; the correlation between X_1 and X_2 is

$$r(X_1, X_2) = a_{11}a_{21}\,\text{var}(F_1) + (a_{11}a_{22} + a_{12}a_{21})\,\text{cov}(F_1, F_2) + a_{12}a_{22}\,\text{var}(F_2)$$
$$= a_{11}a_{21} + a_{12}a_{22} = 0.61.$$

In the general case with m variables, we have an m-dimensional space with an ellipsoid with m mutually perpendicular axes. There are $r \leq m$ eigenvalues larger than 0, in decreasing order. Data reduction is obtained when the data can be represented as parsimonious as possible and without loss of information by less than r principal components. In this way, more insight is obtained into the relationships of the variables. Basic references to PCA are Jolliffe (2002) and Ramsay and Silverman (2002).

8.3 Exploratory factor analysis

A generally accepted definition of factor analysis (FA) is a set of statistical methods of describing the interrelationships of a set of variables by statistically deriving new variables, called factors, that are fewer in number than the original set of variables. Many FA methods have been proposed, and thousands of publications have been written on the subject of FA. Only the basic idea behind FA can be given here; more information on factor analysis can be found in, for example, Gorsuch (1983). We illustrate FA with the principal factor factor analysis or principal axis factor analysis of a correlation matrix, because of its resemblance to PCA.

We start with an m by m correlation matrix. The underlying model is

$$X_i = a_{i1}F_1 + a_{i2}F_2 + \cdots a_{ir}F_r + a_iU_i \tag{8.1}$$

where U is a unique factor plus random error. The factors are standardized and uncorrelated so the variance of variable 1 equals

$$1 = a_{i1}^2 + a_{i2}^2 + \cdots a_{ir}^2 + a_i^2 = h_i^2 + a_i^2 \tag{8.2}$$

where h_i^2 is the communality of variable i, variance based on the common factor space. In FA the variances in the correlation matrix (1), are replaced by estimated communalities. In principal factor factor analysis a PCA is done on the correlation matrix with estimated communalities in the diagonal. The outcome is a matrix with loadings a_{ij}, and with the factors ordered according to the percentage of variance they "explain." One of the problems that confronts the researcher is the number of factors to retain in the final factor solution. One approach to the number-of-factors problem is a visual examination of the decrease of the eigenvalues from an analysis of the original correlation matrix, the scree test. In the scree test one stops adding factors when the decrease in eigenvalues levels off. Another criterion is to stop factoring when the eigenvalues drop below 1. Other procedures can be found in the literature.

The final result of the factor analysis with the principal factor factor analysis remains arbitrary. The obtained factors are orthogonal and ordered according to their contribution to explained variance. There is no reason for this solution to have relevance to the field of investigation. There is a fundamental indeterminacy in factor analysis: the data can be described by a multitude of factor solutions with the same number of factors. We demonstrate this with a factor analysis of ten variables from a study with 24 tests by Holzinger and Swineford (1939), data that have been used by other authors as well. The intercorrelations for the ten tests and their names are given in Table 8.1.

A two-factor solution was deemed to be adequate, convenient because a two-factor solution can adequately be represented in a figure. The factor loadings from the analysis with principal factors are presented in Figure 8.2. All tests load on the first factor. The loadings on the second factor range from positive to negative. The dashed lines represent another possible solution; in fact, the alternative solution was obtained from the original by an orthogonal varimax rotation. Clearly, the alternative representation has an advantage in that some tests load notably on only one factor. A still better solution is obtained with an oblique rotation. With an angle smaller than 90° between the two axes, more tests have small loadings on one of the factors. A positive correlation between factors better reflects the fact that all tests correlate positively. Now, we may with some prudence try to interpret the results. One of the rotated factors, the factor on which the variables 1 though 5 load, corresponds to verbal ability, the other is defined by numerical tests. Series completion loads on both factors. Finally, factor scores can be estimated from the test scores and the FA solution chosen.

Exploratory factor analysis is relevant in, for example, item analysis. With factor analysis one can find out which dimensions underlie the responses to items and which items are good indicators for a certain factor. In this way, meaningful, homogeneous tests can be constructed from a larger pool of items. Linear factor analysis is the subject of this chapter. Nonlinear factor analysis is a subject of the chapter on item response theory.

Table 8.1 Matrix of correlations between ten psychological tests from the Holzinger and Swineford study.

	1	2	3	4	5	6	7	8	9	10
1. General information	1.0									
2. Paragraph comprehension	0.622	1.0								
3. Sentence completion	0.656	0.722	1.0							
4. Word classification	0.578	0.527	0.619	1.0						
5. Word meaning	0.723	0.714	0.685	0.532	1.0					
6. Addition	0.311	0.203	0.246	0.285	0.170	1.0				
7. Counting dots	0.215	0.095	0.181	0.271	0.113	0.585	1.0			
8. Numerical puzzles	0.318	0.263	0.314	0.362	0.266	0.405	0.355	1.0		
9. Series completion	0.435	0.431	0.405	0.501	0.504	0.262	0.350	0.451	1.0	
10. Arithmetic Problems	0.420	0.433	0.437	0.388	0.424	0.531	0.414	0.448	0.434	1.0

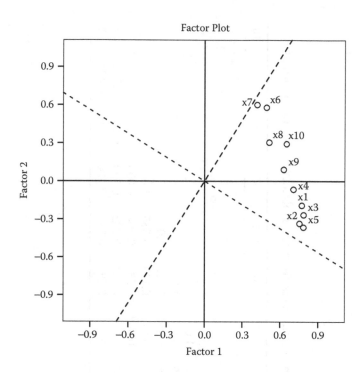

Figure 8.2 A two-factor solution for ten tests from the Holzinger and Swineford study; the dashed lines show the two-factor solution after varimax rotation.

8.4 Confirmatory factor analysis and structural equation modeling

Exploratory factor analysis is an approach to find a model that best fits the correlations or covariances in terms of common factors. The number of common factors and the relationships between these factors and the variables are obtained from the analysis. Confirmatory factor analysis (CFA), in contrast, is used when the researcher has some substantive knowledge such that a structure for factors can be hypothesized and tested (Bryant and Yarnold, 1995), for example, a factor model in which several loadings are set equal to zero. CFA merges into structural equation modeling (SEM), in which models for latent variables and manifest variables are hypothesized and tested.

In SEM and CFA the model can be represented in two ways: by a graphical specification of the model, and by a structural equation

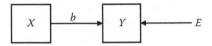

Figure 8.3 Simple regression as a path diagram.

specification of the model. The models are portrayed as path diagrams. These diagrams originate from path analyses used in the pursuit of causal inference with nonexperimental data. Figure 8.3 depicts the path diagram for the regression model $Y = bX + E$. In the figure the measured or observed variables are depicted in rectangles, and the arrows depict the direction of the effect. E is the prediction error (not standardized as in Equation 8.1).

In a model with latent variable, there are two basic assumptions. The first is that the responses on the measured variables are the result of a person's position on the latent variables. The second is that after controlling for the latent variables, the measured variables have nothing in common. The latter assumption is known as the principle of local independence, a principle discussed in more detail in Chapter 9.

Latent variables are represented with a surrounding circle or oval-like element. Figure 8.4 represents a structure with errors and observed

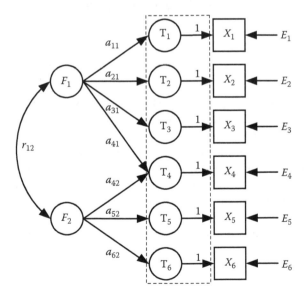

Figure 8.4 A two-factor model with correlated factors and true scores.

scores for six (sub)tests, true scores for these tests, and two underlying common factors F_1 and F_2. The common factors are correlated. From the figure we can obtain the variances and covariances of observed scores:

$$\sigma_4^2 = a_{41}^2 + a_{42}^2 + 2a_{41}a_{42}r_{12} + \sigma_{E_4}^2$$

$$\sigma_{14} = a_{11}a_{41} + a_{11}a_{42}r_{12}$$

and so forth.

There are several softare packages for SEM analyses, like AMOS (www.spss.com, www.assess.com), EQS (www.mvsoft.com), and LIS-REL (www.ssicentral.com). For basic concepts and information on how to build the equations in SEM, the reader is referred to Byrne (1998, 2001, 2006). Statistical model testing with SEM follows a number of steps: check whether the statistical assumptions are satisfied, estimate the model parameters, and evaluate model fit. When a model does not fit, alternative models should be investigated.

SEM plays an important role in test theory. In Chapter 7, for example, the use of SEM in validation research was mentioned. IRT, to be discussed in Chapter 9, has been included in SEM. The interdisciplinary journal *Structural Equation Modeling* reports about the progress in the field.

Exercises

8.1 Perform a PCA on the correlation matrix of Table 8.1. What is the size of the first four components? How much variance is accounted for by the first two principal components? Does it seem reasonable to retain two components?

8.2 Replicate the principal axis solution with two factors. Which subtest has the smallest communality? Relate the communality of this item to the representation of the factor loadings in Figure 8.2.

8.3 Present a SEM model with simple structure for the variance–covariance matrix of the Holzinger and Swineford tests 2, 5, 6, 7, and 9.

CHAPTER 9

Item Response Models

9.1 Introduction

Item response theory is a general term for a family of models, the item
response or IRT models that share some fundamental ideas. These
ideas are that IRT models persons' responses on individual items. The
response of a person on a test item is conceived of as a function of
person characteristics and item characteristics. The response of a
person (i.e., the performance of an examinee) is assumed to depend
upon one or more factors called (latent) traits or abilities. Each item
of a set of items is assumed to measure the underlying trait or traits.
An example of a simple IRT model is that a person's performance on
an item depends only on one underlying trait, and that the relationship
between persons' performance on an item and the trait underlying
item performance can be described by a monotonically increasing function. The latter function is commonly called *item trace line, item characteristic function* (ICF), or *item characteristic curve* (ICC). It specifies
how the probability of a correct response to an item increases as the
level of the trait increases. In contrast to classical test theory and
generalizability theory discussed earlier, IRT consists of a class of
mathematical models for which estimation procedures exist for model
parameters (i.e., person and item parameters) and other statistical
procedures for investigating to what extent the model at hand fits the
data or persons' responses to a set of items.

IRT research and developments not only pervade scholarly journals, in the latest edition of the *Standards for Educational and Psychological Testing* (APA et al., 1999), ample space is given to IRT.

In Section 9.2 the basic concepts of IRT will be discussed, and
several unidimensional models for dichotomous data will be introduced. Apart from the types of IRT models in terms of a specification
of the ICC, models can also be distinguished as to the number of
response options modeled, and also more than one latent trait can be

postulated, leading to multidimensional IRT models (see also Section 9.3). In Section 9.4 item-test regression will be considered and compared to IRT item-trait regression. It has already been said that IRT leads to the estimation of model parameters; the estimation of item parameters is introduced in Section 9.5. In Section 9.6 the joint maximum likelihood (JML) estimation procedure for item as well as person parameters is discussed. To what JML leads in the Rasch model can be found in Section 9.7. Other estimation methods with their characteristic properties will be discussed in Sections 9.8 through 9.10. In Section 9.11 some specific problems will be discussed with respect to the estimation of item parameters. Section 9.12 is on maximum likelihood (ML) estimation of person parameters. This does not exhaust the possibilities for the estimation of person parameters. In Section 9.13 Bayesian estimation is mentioned.

The IRT concepts of item information and test information break away from the concept of the variance of measurement errors being constant over the whole range of scores. These information concepts are elaborated in Section 9.14. As IRT gives a statistical model approach to measurement, model fit is also a central theme (Section 9.15). Finally, for the interested reader, maximum likelihood estimation in the context of the Rasch model is discussed in an appendix (Section 9.16).

9.2 Basic concepts

In a unidimensional model, we assume that the responses can be described by a model with one latent dimension (i.e., as if only one dimension ability or, more generally, latent trait accounts for the responses). The latent ability θ is defined in most models on a scale of minus infinity to plus infinity $(-\infty, \infty)$. The probability of a particular response to a dichotomous item is a monotonous and nonlinear function of the ability. The probability of a correct response increases with increasing ability or latent trait value (for an exception in connection with a model involving guessing behavior see Samejima, 1979). The conditional probability correct, the probability of a correct response given ability, might be interpreted as the probability of a correct response for a randomly selected person with the given ability (Holland, 1990).

The assumption of *unidimensionality* implies that the responses to different items are independent given the latent trait. We have *local independence*. If we did not have local independence, one dimension

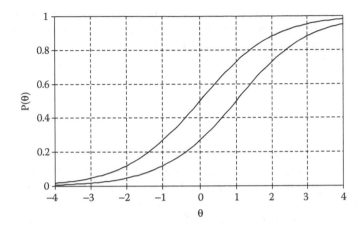

Figure 9.1 Item characteristic curves for two Rasch items.

would not be enough to account for the responses. In that case, local independence would be obtained given all relevant latent traits.

9.2.1   The Rasch model

In Figure 9.1 the probability of a correct response as a function of latent ability θ is given for two hypothetical items. The two items are Rasch items (i.e., they satisfy the Rasch model assumption of equal discriminability for items with correct–incorrect scoring). That is to say, Rasch curves are parallel.

In the Rasch model, the probability of a correct response on item i, given ability or person parameter θ, is equal to

$$P_i(\theta) = P(x_i = 1 \mid \theta) = \frac{e^{(\theta - b_i)}}{1 + e^{(\theta - b_i)}} = \frac{\exp(\theta - b_i)}{1 + \exp(\theta - b_i)} \tag{9.1}$$

where b_i is the item difficulty parameter of item i (Rasch, 1960). The probability varies from 0.0 for $\theta = -\infty$ to 1.0 for $\theta = \infty$. Most of the variation in probability lies in the interval from $\theta = b_i - 4.0$ to $\theta = b_i + 4.0$. For θ equal to b_i the probability equals one half. In Figure 9.1 the curve on the left is the curve of a Rasch item with $b_i = 0$; the curve on the right belongs to a Rasch item with $b_i = 1.0$.

The person parameter θ and the item parameter b appear in Equation 9.1 only in the combination $\theta - b$. When we take log-odds (i.e., the natural logarithm of the ratio of the probability of a correct response and the probability of an incorrect response), we obtain the logit:

$$\ln\left(\frac{P_i(\theta)}{1 - P_i(\theta)}\right) = \theta - b_i$$

We conclude from this that the probability of a correct response (Equation 9.1) remains invariant if we increase the value of θ to $\theta^* = \theta + d$ and simultaneously increase the value of b to $b^* = b + d$. The parameters of the Rasch model are defined on an additive scale, a special case of the interval scale. The consequence is that in an application of the model, one restriction on the parameters is always needed to fix the latent scale. We might, for example, set the item parameter of one of the items equal to 0.0. Another possible restriction is to set the mean of the item parameters equal to 0.0.

9.2.2   Two- and three-parameter logistic models

The Rasch model is a one-parameter model. The model has one item parameter: the item difficulty parameter. In many tests the items differ not only with respect to difficulty but also with respect to discriminating power. The two-parameter *logistic model* (Birnbaum, 1968), 2PL model for short, has a second item parameter, item discrimination. The model is given by the following equation:

$$P_i(\theta) = \frac{\exp[a_i(\theta - b_i)]}{1 + \exp[a_i(\theta - b_i)]} \tag{9.2}$$

where a_i is the discrimination parameter of item i. The term *logistic* refers to the fact that the right-hand side of Equation 9.2 is equal to the cumulative logistic distribution function. The slope of the ICC at θ is equal to $a_i P_i(\theta)[1 - P_i(\theta)]$. At $\theta = b_i$ the slope is equal to $0.25a_i$. So the slope at $\theta = b_i$ is steeper for higher values of a_i.

The person parameters in the 2PL model are defined on an interval scale. The probability of a correct response does not change if we transform θ into $\theta^* = d\theta + e$ under simultaneous transformations $b^* = db + e$ and $a^* = a/d$. In order to fix the latent scale, we need two restrictions, for example, the mean θ can be set equal to 0.0 and the standard deviation of the θ's can be set equal to 1.0.

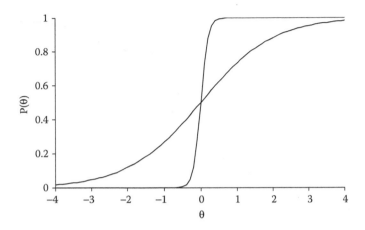

Figure 9.2 Item characteristic curves for two items with different discrimination parameters.

In Figure 9.2 two item characteristic curves are displayed: one with $a_i = 1.0$, the other with $a_i = 10.0$. One can imagine what will happen to the ICC if the discrimination parameter of an item increases indefinitely. The item characteristic curve approximates a jump function with a value equal to 0 for θ smaller than b_i, and a value equal to 1 for θ larger than b_i. We then have a Guttman item with a perfect discrimination at the value $\theta = b_i$, and no discriminating power to the left and to the right of this point.

The item characteristic curves in Figure 9.2 cross. In the Rasch model, item characteristic curves do not cross, but run parallel. The Rasch model is not the only probabilistic model with nonintersecting item characteristic curves. Another model with this property is the nonparametric Mokken model of double monotonicity (Mokken, 1971).

With items of the multiple-choice type, guessing is possible and cannot be excluded. If an examinee does not know the answer on a four-choice item, he or she might correctly guess the answer with a probability equal to one fourth. With this kind of item, one better introduces a lower asymptote larger than 0 for the item characteristic curve. The three-parameter logistic model is obtained:

$$P_i(\theta) = c_i + (1 - c_i) \frac{\exp[a_i(\theta - b_i)]}{1 + \exp[a_i(\theta - b_i)]} \qquad (9.3)$$

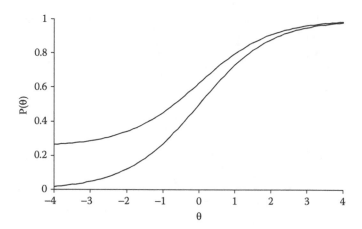

Figure 9.3 Item characteristic curves for two items with different pseudo-chance-level parameters.

where c_i is the lower asymptote. The third parameter is called the *pseudo-chance-level parameter*. This parameter is not set equal to the inverse of the number of response alternatives, but it is estimated along with the other item parameters. Figure 9.3 displays two items: one with c_i equal to one fourth, and the other with c_i equal to 0.0. The influence of the third item parameter at the lower level of θ is clear.

The 2PL model is a special case of the 3PL model with $c_i = 0.0$ for all items. The 1PL model is obtained if all item discrimination parameters are set equal. In the Rasch model (Equation 9.1), this common discrimination parameter is set equal to 1.0. The differences between the models seem to be very clear. Meredith and Kearns (1973), however, demonstrated that a special case of the 3PL model can be reformulated in terms of the Rasch model.

In addition to the logistic model, we have the *normal ogive model*. It has an ICC with the form of the cumulative normal distribution. This model was the first to be used in test theory (see Lord, 1952). The two-parameter normal ogive model is given by

$$P_i(\theta) = \Phi[a_i(\theta - b_i)] = \int_{-\infty}^{a_i(\theta-b_i)} \varphi(t)dt = \int_{-\infty}^{a_i(\theta-b_i)} \frac{1}{\sqrt{2\pi}} \exp\left[-\frac{1}{2}t^2\right]dt \quad (9.4)$$

The normal ogive plays a role in some models with more dimensions or more than two response categories (Bock, Gibbons, and Muraki, 1988;

Muraki and Carlson, 1995; Muthén, 1984). The application of the model
to polytomous, multidimensional data will be discussed in Section 9.3.

The normal ogive model and the logistic model give practically the
same probabilities if a scaling factor $D = 1.7$ is introduced in the logistic
model:

$$\Phi[a_i(\theta - b_i)] \approx \frac{\exp[Da_i(\theta - b_i)]}{1 + \exp[Da_i(\theta - b_i)]}$$

For this reason, the factor D is frequently part of the logistic model
as described in the literature. If the parameters of the logistic model
are given by Equation 9.2 or Equation 9.3, the parameters are defined
in the "logistic metric." They can be transformed to the "normal metric"
through a division of the discrimination parameters by the scaling
factor 1.7.

9.2.3   Other IRT models

Fischer (1983) extended the Rasch model with linear constraints on
the item parameters. Mislevy (1983) applied IRT to grouped data.
Patz, Junker, Johnson, and Marino (2002) suggested an IRT approach
for constructed response items where answers are rated by several
raters.

Other extensions of the models have been proposed: extensions to
more than two response categories and extensions to more than one
latent trait dimension. Bock (1972) proposed a general *nominal
response model*. In this model the probability of a response to item
option k from m available options is given by

$$P_{ik}(\theta) = \frac{\exp[a_{ik}(\theta - b_{ik})]}{\sum_{h=1}^{m} \exp[a_{ih}(\theta - b_{ih})]} \qquad (9.5)$$

Extensions to items with more than two categories have been
formulated notably for the Rasch model (e.g., Andersen, 1977). Two
closely related Rasch models for items with more than two categories
are well known: the *rating scale model* (Andrich, 1978, 1999) and the
partial credit model (Masters, 1982, 1999). The rating scale model can
be obtained as a submodel of the partial credit model by a reparame-
trization of the parameters. The models differ in their substantive
interpretations, however.

The partial credit model models partial understanding in problems with multiple steps. In the partial credit model with categories 0,..., m, the probabilities for categories other than category 0 are given by

$$P_{ik}(\theta) = \frac{\exp\left[\sum_{j=1}^{k}(\theta - \delta_{ij})\right]}{1 + \sum_{h=1}^{m}\exp\left[\sum_{j=1}^{h}(\theta - \delta_{ij})\right]} \qquad (9.6)$$

The model parameters δ_{ij} are "step" difficulties governing the "step" probabilities $P_{ik}/(P_{ik} + P_{i,k-1})$, which have the form of Rasch model items. The model is, for example, applied when several dichotomous items are related and form an item cluster. Huynh (1994) demonstrated the applicability of the PCM model for a testlet composed of independent Rasch items. When the steps of an item are scored sequentially (all steps after the first error in a chain of steps are evaluated as failed), the model is not suitable. Sequentially scored items must be modeled by a set of binary items with missing observations after the first failure (Akkermans, 2000).

In the rating scale model, thresholds are modeled for the different categories in items with ordinal response categories (for example, *poor*, *fair*, *good*, *excellent*). In the rating scale model for $m + 1$ score categories $(0,..., m)$, the probability of choosing category k of item i can be written as

$$P_{ik}(\theta) = \frac{\exp\left[k(\theta - b_i) - \sum_{j=0}^{k}\tau_{j(m+1)}\right]}{\sum_{h=0}^{m}\exp\left[h(\theta - b_i) - \sum_{j=0}^{h}\tau_{j(m+1)}\right]} \qquad (9.7)$$

where $\tau_{j(m+1)}$ are threshold parameters for all items with a common number of $m + 1$ categories, with $\tau_{0(m+1)} = 0$.

An exemplary rating scale item with five categories, $b = 0.0$, and threshold parameters -1.8, -0.8, 0.8, and 1.8 is given in Figure 9.4. As can be seen from Figure 9.4a, the option probabilities for categories j and $j - 1$ are equal at $\theta = \tau_j + b$.

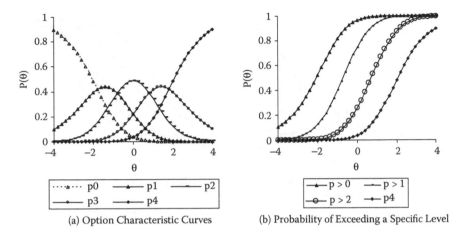

(a) Option Characteristic Curves   (b) Probability of Exceeding a Specific Level

Figure 9.4 The rating scale model.

Samejima (1969) published on a *graded response model* for items with ordered response options. For the logistic model the probability of choosing option k or a higher option equals

$$P^*_{ik}(\theta) = \frac{\exp[a_i(\theta - b_{ik})]}{1 + \exp[a_i(\theta - b_{ik})]} \tag{9.8}$$

where the option parameter b_{ik} increases with k. In this model, the probability of option k as a function of latent ability, the option characteristic function, is

$$P_{ik}(\theta) = P^*_{ik}(\theta) - P^*_{i,k+1}(\theta) \tag{9.9}$$

with $P^*_{i0}(\theta) = 1.0$, $P^*_{i,m+1}(\theta) = 0.0$. An example of an item in the graded response model is given in Figure 9.5 (with parameters $a = 1.0$, $b_1 = -2.0$, $b_2 = -0.75$, $b_3 = 0.75$ and $b_4 = 2.0$). The probability of category 0 decreases with θ, the probability of category m increases with θ, and the probabilities of the intermediate categories have their peaks in the order of the categories. A submodel of this model is the model where b_{ij} can be partitioned into an item parameter b_i and a scale parameter τ_j (Muraki, 1990).

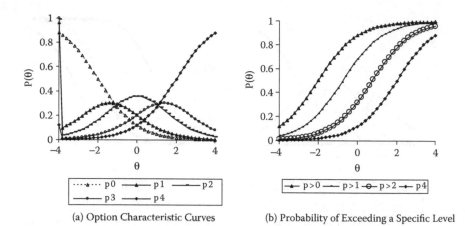

(a) Option Characteristic Curves   (b) Probability of Exceeding a Specific Level

Figure 9.5 The graded response model.

In the graded response model, the probability of a response in category k or $k + 1$ is

$$P_{ik}(\theta) + P_{i,k+1}(\theta) = [P_{ik}^*(\theta) - P_{i,k+1}^*(\theta)] + [P_{i,k+1}^*(\theta) - P_{i,k+2}^*(\theta)]$$

$$= P_{ik}^*(\theta) - P_{i,k+2}^*(\theta)$$

which has the same form as the original probabilities. The response categories *poor* and *fair*, and the response categories *good* and *excellent* of the four response categories *poor, fair, good, excellent* might, for example, be combined for the analysis of the response data. The graded response model allows a dichotomization of the response categories, and—at the cost of losing some information—an analysis with an IRT model for dichotomous data. Combining categories is not possible with the Rasch models for polytomous data without violation of the model (Jansen and Roskam, 1986). So, in case the combining of categories is an obvious possibility with the data that are to be analyzed with an IRT model, the graded response model is the more realistic model despite the statistical advantages of the polytomous Rasch models. However, the partial credit model (or the equivalent rating scale model) might give comparable results.

Molenaar presented a generalization of the Mokken model to the case of more than two categories (Molenaar, 1997). Rossi, Wang, and Ramsay (2002) also suggested nonparametric IRT modeling. An introduction to nonparametric modeling is given by Molenaar and Sijtsma (2002). Rasch developed a Poisson model (see also Lord and Novick, 1968) for the number of mistakes in a test. Models for speeded and time-limit tests can be found in Roskam (1997).

The other extension is to more dimensions. Item response models are actually factor-analytic models with a nonlinear relationship between factor and expected scores. Most models proposed are compensatory: a low-ability i can be compensated by a high-ability j (like in Equation 8.1). A factor-analytic approach to a multidimensional latent space was given by Bock, Gibbons, and Muraki (1988), McDonald (1997, 1999) with NOHARM (Fraser, 1988), Muraki and Carlson (1995), Reckase (1997), and Muthén (1984) with LISCOMP (and Mplus, see Muthén, 2002; www.statmodel.com). Shi and Lee (1997) discussed the estimation of latent abilities for the nonlinear factor model. For multidimensional Rasch models, see Adams, Wilson, and Wang (1997). De la Torre and Patz (2005) discuss a special case of multidimensional IRT in which the items exhibit a simple structure (i.e., each item loads on one latent trait only). This situation may arise when a test battery with several subtests is administered in one test session. A noncompensatory multidimensional model in which cognitive processes are modeled was proposed by Embretson (1984).

9.3 The multivariate normal distribution and polytomous items

In the factor analytic model, z_{ip}, the observed score of person p on item i, can be written as a weighted sum of factor scores plus error:

$$Z_{ip} = \alpha_{i1}\theta_{1p} + \alpha_{i2}\theta_{2p} + \cdots + \alpha_{in}\theta_{np} + E_{ip} \qquad (9.10)$$

In this equation, the additive constant was dropped. This can be done without consequences for the generality of our argument.

In this chapter, the Z_{ip} are not observed; they describe response processes. When Z_{ip} exceeds a certain threshold, a score in a given category of polytomous item i is observed. For item i with $m + 1$ categories, we have

$$\begin{aligned}
X_{ip} &= 0 \quad \text{if } Z_{ip} < \gamma_{i0} \\
X_{ip} &= k \quad \text{if } \gamma_{i,k-1} \le Z_{ip} < \gamma_{ik} \\
X_{ip} &= m \quad \text{if } \gamma_{i,m-1} < Z_{ip}
\end{aligned} \qquad (9.11)$$

Let us write $P^*_{ik}(\theta)$ for the probability $P(Z_i \geq \gamma_{i,k-1} \mid \theta_1,...,\theta_n)$. Then the probability of a response in category k can be written as

$$P_{ik}(\theta) = P^*_{ik}(\theta) - P^*_{i,k+1}(\theta) \tag{9.12}$$

a generalization of the graded response model, which was introduced for one-dimensional latent space.

The next step is to define the distribution of errors E_{ip}. In this section we assume that the errors are normally distributed with variance σ_i^2. It follows that

$$P^*_{ik}(\theta) = \Phi\left(\frac{\sum_{j=1}^{n}\alpha_{ij}\theta_j - \gamma_{i,k-1}}{\sigma_i}\right) \tag{9.13}$$

For a one-dimensional model, the probability $P^*_{ik}(\theta)$ is a normal ogive curve. The relation between Equation 9.13 and the normal ogive model for dichotomous data (Equation 9.4) is discussed in Exhibit 9.1.

Exhibit 9.1   The relationship between model 9.13 and the normal ogive model for dichotomous data and one-dimensional latent space

What is the relationship between the model discussed in this section and Model 9.4 for a one-dimensional latent space and dichotomously scored items?

With dichotomous data, we can write the threshold model with normally distributed errors as

$$Z_{ip} = \theta_p + E_{ip}$$

$$X_{ip} = 0 \quad \text{if} \quad Z_{ip} < b_i$$
$$X_{ip} = 1 \quad \text{if} \quad Z_{ip} \geq b_i$$

and

$$P_i(\theta) = \Phi\left(\frac{\theta - b_i}{\sigma_i}\right) = \Phi[a_i(\theta - b_i)]$$

where

$$a_i = 1/\sigma_i$$

Without loss of generality, we can assume that θ has a mean equal to zero and a variance equal to one. The variance of Z_i is equal to

$$\sigma_{Z_i}^2 = \sigma_\theta^2 + \sigma_i^2 = 1 + a_i^{-2}$$

The correlation between Z_i and θ is

$$\rho_i = \rho_{\theta Z_i} = \frac{\sigma_\theta^2}{\sigma_\theta \sqrt{\sigma_\theta^2 + \sigma_i^2}} = \frac{a_i}{\sqrt{1 + a_i^2}}$$

Let us now assume that θ is normally distributed. The correlation ρ_i can be approximated by the biserial correlation between the item and the total test or rest-test, if the test is long. The biserial correlation estimates the correlation of a normally distributed variable assumed to underlie the dichotomous scores on a variable, with a continuous variable. The proportion correct of item i is equal to

$$\pi_i = \Phi\left(\frac{-b_i}{\sqrt{1 + a_i^{-2}}}\right)$$

It is clear that the item parameters a_i and b_i can be obtained from ρ_i and π_i.

Now, let the latent variables as well as the item scores be standardized; the thresholds are assumed to be defined in this metric. If we assume that the latent variables θ have a multivariate normal distribution, the response processes Z are multivariate normally distributed. The marginal distribution of Z_i is the standard normal distribution. Then

$$P(Z_i \geq \gamma_{i,k-1}) = \Phi(-\gamma_{i,k-1})$$

Also, the joint distribution of any two variables Z_i and Z_j has the shape of a bivariate normal distribution. If the variables Z had been observed, we would have used linear factor analysis to obtain the item parameters α. Now we have the joint frequencies for pairs of variables X_i and X_j.

The marginal frequencies of the variables X_i and the joint frequencies for pairs of variables can be used for the estimation of the factor loadings and the thresholds (Muthén, 1984). The maximum likelihood approach, which is discussed in more detail in this chapter, can also be used for the estimation of the item parameters. The maximum likelihood approach (Muraki and Carlson, 1995) uses the information of all answer patterns and, therefore, is called full-information factor analysis.

The model does not have a pseudo-guessing parameter. In case guessing plays a role, some adaptations of the procedure are necessary. Bock et al. (1988) apply full-information factor analysis to dichotomous data. They discuss the effect of guessing and present analyses of a section of the LSAT uncorrected and corrected for guessing.

9.4 Item-test regression and item response models

Consider Figure 9.6. It depicts the item-test regressions for two items of a ten-item test. The regressions do not correspond to fixed item characteristics. We would have obtained different regressions with different total tests and different examinee groups.

Let us now look at Figure 9.7. We have the same items as in Figure 9.6. However, the abscissa is redefined. We have, for example, enlarged the difference between "8 correct" and "9 correct" in comparison to the difference between "5 correct" and "6 correct" (the first interval is nearly a factor 1.7 larger than the second interval). We chose different units and relabeled the axis as the latent trait dimension θ. In addition, we fitted curves to the empirical regressions. The curves fit adequately. The curves are identical apart from a translation along the horizontal axis.

How did we proceed? We fitted the Rasch model to the data from the ten-item test. We used an estimation method in which the transformed total score is used as a proxy to θ. Is it correct to state that

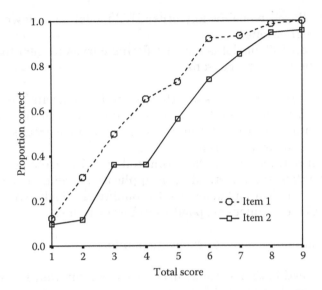

Figure 9.6 Item-test regression for two items from a ten-item test.

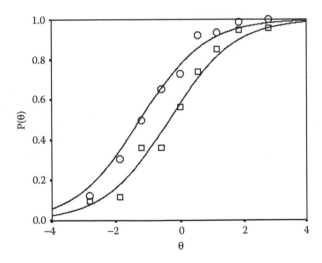

Figure 9.7 Estimated item-trait regressions (ICCs) for the two items from Figure 9.6.

fitting the Rasch model or any other IRT model boils down to fitting
item-test regressions?

Fitting an IRT model is not just fitting curves to item-test regressions for the following reasons:

- The item parameters are thought to be invariant item characteristics for the relevant population. The same ICCs should
  be obtained with different tests and examinee groups because
  of the invariance property.
- Item parameters can be estimated even in case no examinee
  takes the same test. Also, in applications we do not have to
  think in terms of fixed tests. In computerized adaptive testing
  (CAT), examinees respond to different items.

We mention some other relevant points:

- The estimation method used does not minimize a sum of
  weighted squared differences between the empirical regression and the curve. But such a weighted sum is used in one
  approach to determine item fit (see Section 9.15).
- The total score is used only for estimation in the Rasch model
  (i.e., the total score is a sufficient statistics for the estimation
  of in the Rasch model).
- Determining item fit is just one aspect of model fit. We must
  assess, for example, whether the data are unidimensional.
  Within a given score group, the responses to two items should
  be independent. (To be more precise, we expect a small negative correlation between the responses due to the fact that
  the item scores within a score group sum to a constant value.)
- The estimation method used (JML, see Section 9.7) is statistically faulty. Examinees with the same total score do not have
  the same value for θ. As a result of the incorrect simplification
  in JML, the item parameter estimates are biased and the ICCs
  in the figure are farther apart than they should be.

9.5 Estimation of item parameters

In a model with one latent trait, the response probabilities given the
latent trait are locally independent, as defined in Equation 6.16. In
case the responses are known, we call Equation 6.16 the likelihood of
the given response pattern instead of the probability. The likelihood

of the responses x_{ip} of N persons ($p = 1,..., N$) on n dichotomous items ($i = 1,..., n$) is given by

$$L = \prod_{p=1}^{N}\left(\prod_{i=1}^{n}P(X_{ip} = x_{ip} | \theta_p)\right) = \prod_{p=1}^{N}\prod_{i=1}^{n}P_i(\theta_p)^{x_{ip}}[1 - P_i(\theta_p)]^{1-x_{ip}} \qquad (9.14)$$

where $x_{ip} = 1$ for a correct response, and $x_{ip} = 0$ for an incorrect response.

The main method for the estimation of the item parameters is the maximum likelihood method. There are four alternative maximum likelihood methods:

- In joint maximum likelihood (JML) estimation, person and item parameters are estimated jointly.
- In marginal maximum likelihood (MML) estimation, the person parameters are eliminated from the estimation process by integration over the distribution of person parameters.
- In Markov chain Monte Carlo (MCMC) estimation, parameter estimates as well as posterior distributions of parameters are obtained through a sampling approach. MCMC is very suitable for estimation in the context of complex (hierarchical) models.
- In conditional maximum likelihood (CML) estimation, the person parameters are eliminated from the estimation process by conditioning on the total scores. The method is available for the logistic model with only a difficulty parameter. So, CML is possible with the Rasch model. Andersen (1983; see also Verhelst and Glas, 1995) demonstrated that CML is also possible when in the two-parameter model the slopes are assumed to be known.

Each of these methods will be discussed: JML in Sections 9.6 and 9.7, MML in Section 9.8, MCMC in Section 9.9, and CML in Section 9.10. More information on estimation techniques is given in Baker and Kim (2004).

Cohen (1979; see also Wright and Stone, 1979) proposed a simple approximation for the estimation of parameters in the Rasch model, assuming normally distributed item and person parameters. His approximation is based on the similarity between the logistic model and the cumulative normal distribution function. Urry (1974) suggested to estimate item parameters in the three-parameter normal ogive model from item indexes under the assumption of normally distributed person parameters (see also Lord and Novick, 1968). The approximations might produce good starting estimates of parameters for the likelihood procedures.

For the multivariate normal distribution of latent abilities, not only MML has been proposed, but also the analysis of marginals for items and item pairs (see Section 9.3).

9.6 Joint maximum likelihood estimation for item and person parameters

In the JML method for dichotomous data, person and item parameter estimates are obtained that maximize the likelihood in Equation 9.14. First, starting values for the parameters are computed. Then, person parameters are computed that maximize the likelihood given the item parameter estimates. Next, new item parameter estimates are obtained on the basis of the current estimates of the person parameters. One cycles through this process until the estimates from both sets of parameters are stable (i.e., until the changes in the estimates fall below a threshold).

Actually the natural logarithm of the likelihood, the log likelihood, is maximized. Maximizing the log likelihood is easier than maximizing Equation 9.14, and it produces the same estimates. Before the process is started, persons with 0 responses correct and persons with perfect scores are eliminated; the maximum likelihood estimate of θ is $-\infty$ for a total score equal to 0 and ∞ for a total score equal to the number of items n. On similar grounds, items that are answered correctly by all persons or by no persons are removed.

The estimation procedure makes use of the fact that at its maximum a function has a zero slope. So, new parameter estimates are obtained by taking derivatives of the log likelihood with respect to these parameters, setting the results equal to zero, and solving for the parameters (see Section 9.16). The estimation equation for person parameter θ_p, the equation from which the new estimate of θ_p is to be obtained, can be written as

$$\sum_{i=1}^{n} \frac{x_{ip} - P_i(\theta_p)}{P_i(\theta_p)[1 - P_i(\theta_p)]} \frac{\partial P_i(\theta_p)}{\partial \theta_p} = 0 \qquad (9.15)$$

where $\partial P_i(\theta_p)/\partial \theta_p$ is the derivative of $P_i(\theta_p)$ with respect to θ_p.

The estimation equation for an item parameter of item i is

$$\sum_{p=1}^{N} \frac{x_{ip} - P_i(\theta_p)}{P_i(\theta_p)[1 - P_i(\theta_p)]} \frac{\partial P_i(\theta_p)}{\partial \gamma_i} = 0 \qquad (9.16)$$

where γ_i is the item parameter of item i in question.

There are N estimation equations for person parameters and n (Rasch model), $2n$ (2PL model), or $3n$ (3PL model) equations for item parameters. There is also one scale restriction (Rasch model) or two scale restrictions (2PL and 3PL model).

The estimation of parameters in the 2PL model and the 3PL model is not free of problems. In the 3PL model there are response patterns for which no unique maximum for θ exists (Samejima, 1973). Estimation of the lower asymptote c_i might also give problems. In the 2PL model and the 3PL model, the a_i-parameter estimates might be unstable. For that reason, the change in parameter estimates from one iteration to the other is restricted in estimation programs. One way of restraining change is to introduce prior distributions for the parameters (Swaminathan and Gifford, 1986).

9.7 Joint maximum likelihood estimation and the Rasch model

The Rasch model is the simplest model and for this reason very suitable for the introduction of JML. For the Rasch model, the sets of equations for person parameters (Equation 9.15) and item parameters (Equation 9.16) can be simplified to

$$t_p = \sum_{i=1}^{n} P_i(\theta_p), \quad p = 1, \dots, N \qquad (9.17)$$

where t_p is the total score of person p, and

$$s_i = \sum_{p=1}^{N} P_i(\theta_p), \quad i = 1, \dots, n \qquad (9.18)$$

where s_i is the total number of correct responses to item i. In Equation 9.17, the total score of a person is set equal to its expected value; in Equation 9.18, the item total score is set equal to its expectation. Equations 9.17 and 9.18 have to be solved iteratively for the parameters. One restriction must be added in order to fix the additive scale of the Rasch model.

From Equation 9.17 it follows that all persons with the same total score have the same estimated θ. The total score is a sufficient statistic, a result that is valid only for the Rasch model. The implication is that JML estimation for the Rasch model can be viewed as a logistic regression problem with total scores and items as levels of two categorical variables.

In JML estimation, the item parameter estimates depend on the person parameter estimates and vice versa. This causes an estimation problem of biased parameter estimates. The bias in the item parameters does not disappear if the number of persons N increases. With large tests the bias is negligible (see Exhibit 9.2).

Exhibit 9.2   Where JML goes wrong

Suppose we have one latent trait value, with $\theta = 0.0$, and two Rasch items, with $b_1 = 0.0$ and $b_2 = 0.5$.

The probability of a correct response to the items is $p_1 = P_1(\theta = 0) = 0.5$ and $p_2 = P_2(\theta = 0) = 0.37754$.

We obtain the following probabilities:

$P(1 \text{ correct}; 2 \text{ incorrect}) = p_1(1 - p_2) = 0.31123$
$P(1 \text{ incorrect}, 2 \text{ correct}) = p_2(1 - p_1) = \underline{0.18877}$
$P(t(\text{total score}) = 1)$ $= 0.5$
$P(1 \text{ correct} \mid t = 1) = P(1 \text{ correct}; 2 \text{ incorrect})/P(t = 1) = 0.62246$
$P(2 \text{ correct} \mid t = 1) = P(1 \text{ incorrect}; 2 \text{ correct})/ P(t = 1) = 0.37754$

In JML we estimate the item parameters from the probabilities correct given the estimates of abilities. With two items we have only one ability estimate, for $t = 1$; this ability estimate can be set equal to 0.0 (which fixes the latent scale). We solve b_1 from $P(1 \text{ correct} \mid t = 1)$, and b_2 from $P(2 \text{ correct} \mid t = 1)$:

$$\hat{b}_1 = -\ln\left(\frac{P(1 \text{ correct} \mid t = 1)}{1 - P(1 \text{ correct} \mid t = 1)} \right) = -0.5$$

and

$$\hat{b}_2 = -\ln\left(\frac{P(2 \text{ correct} \mid t = 1)}{1 - P(2 \text{ correct} \mid t = 1)}\right) = 0.5$$

The difference between the estimated b parameters is twice the true difference. The problem with the JML method is that it incorrectly equates the empirical regressions $P(i \text{ correct} \mid t)$ with $P_i(\theta)$.

Andersen (1972) demonstrated that with two Rasch items the difference between the JML estimates is always twice the true value. For a test with $n > 2$ Rasch items, a correction factor $(n - 1)/n$ seems adequate for all practical purposes (Wright, 1988).

9.8 Marginal maximum likelihood estimation

In MML person parameters are eliminated by integration over a distribution of θ. The mean of this distribution is set equal to 0.0. The standard deviation of the distribution is set equal to 1.0. This is permitted for the 2PL and the 3PL models, models defined on an interval scale. So, for a Rasch analysis, we start with the 1PL model (i.e., the 2PL model with a common discrimination parameter). The final MML difficulty parameter estimates are transformed to the scale with a common discrimination parameter equal to 1.0.

We do not know the population distribution of person parameters. We have to make some assumptions with respect to the distributional form. The choice we make has an effect on the values of the item parameter estimates. This dependency seems to be the weak spot in MML. Fortunately, it is possible to estimate properties of the latent distribution along with the item parameters when enough data are available.

In case we do not have any information on the distribution of person parameters, a natural choice is a more or less bell-shaped distribution. Traditionally the obvious choice is the normal distribution. A disadvantage is, however, that the distribution leads to awkward computations. This problem can be overcome—we can approximate the normal distribution by a discrete distribution with any degree of accuracy.

Let us consider a discrete distribution of θ with q latent classes. The relative frequencies of θ_k ($k = 1,...,q$) are denoted by $g(\theta_k)$. When the values θ_k and $g(\theta_k)$ are chosen for an optimal approximation of a continuous distribution, they are called quadrature points and quadrature weights, respectively.

The probability of response pattern $\mathbf{x} = (x_1, ..., x_i, ..., x_n)$ given θ_k is written as $P(\mathbf{x} \mid \theta_k)$. If a randomly chosen person from the population makes the n-item test, the probability of response pattern $\mathbf{x}$ equals

$$P(\mathbf{x}) = \sum_{k=1}^{q} P(\mathbf{x} \mid \theta_k) g(\theta_k) \qquad (9.19)$$

Now we administer the test. S response patterns occur. Response pattern l ($l = 1, ..., S$) occurs r_l times.

In MML we maximize the natural logarithm of

$$L_{\text{MML}} = C \prod_{l=1}^{S} P(\mathbf{x}_l)^{r_l} \qquad (9.20)$$

where C is independent of the parameters (Bock and Aitkin, 1981). The resulting estimation equations can be written as

$$\sum_{k=1}^{q} \frac{n_{ik} - n_k P_i(\theta_k)}{P_i(\theta_k)[1 - P_i(\theta_k)]} \frac{\partial P_i(\theta_k)}{\partial \gamma_i} = 0 \qquad (9.21)$$

where γ_i is one of the item parameters of item i, n_{ik} is the posterior expectation of the number correct on the item in latent class k, and n_k is the posterior "size" of latent class k.

In the EM algorithm, we solve iteratively for maximum likelihood estimates and update values n_{ik} and n_k in the "Expectation" step until convergence is reached.

9.9 Markov chain Monte Carlo

MCMC resembles JML in as far as in both approaches item and person parameters are estimated together. MCMC, however, is based on the use of prior distributions for all parameters. MCMC estimation makes use of posterior distributions.

The posterior distribution of θ_i can be written as

$$p(\theta_i \mid \boldsymbol{\beta}, \theta_1, \ldots, \theta_{i-1}, \theta_{i+1}, \ldots, \theta_N, \mathbf{X}) = Cp(\mathbf{X} \mid \boldsymbol{\theta}, \boldsymbol{\beta}) p(\boldsymbol{\theta}, \boldsymbol{\beta}) \qquad (9.22)$$

where $\mathbf{X}$ is the data matrix, $\boldsymbol{\theta}$ is the vector with N person parameters, $\boldsymbol{\beta}$ is the vector with all item parameters, and C is the normalizing constant. When we sample from the posterior distribution, we can use the sample data in order to compute characteristics of this distribution (e.g., the posterior mean and standard deviation). The sampling process can also be used to obtain the final person parameter estimates as well as the final item parameter estimates. This is done using a Markov chain Monte Carlo technique (Gelman, Carlin, Stern, and Rubin, 2004).

We begin with starting values for item and person parameters. Next we sample new parameters from the posterior distributions. This process is repeated. In cycle $k + 1$ we draw, for example, a new value θ_i from the distribution:

$$p(\theta_i^{k+1} \mid \boldsymbol{\beta}^k, \theta_1^{k+1}, \ldots, \theta_{i-1}^{k+1}, \theta_{i+1}^k, \ldots, \theta_N^k, \mathbf{X})$$

At the end of the cycle, all parameter estimates have been updated. When the process stabilizes (i.e., when the posterior distribution does not change with new cycles), sampling is continued in order to obtain the aforementioned characteristics of the posterior distributions. When the person parameters are estimated this way as posterior means, the estimates for equal answer patterns can differ slightly due to sampling variation (Kim, 2001).

The procedure described here is known as the Gibbs sampler. The Gibbs sampler requires the computation of the normalizing constant. This may not be an easy task. Sometimes one works around this problem by a process called data augmentation in which missing data are created (Albert, 1992). Other MCMC methods have been developed that do not depend on the computation of the normalizing constant. An overview of MCMC methods in the context of IRT modeling is given by Patz and Junker (1999). Many researchers use the BUGS software for MCMC estimation (Spiegelhalter, Thomas, Best, and Lunn, 2003). Sinharay, Johnson, and Stern (2006) discuss model checking with the Bayesian posterior predictive model-checking method.

9.10 Conditional maximum likelihood estimation in the Rasch model

In the Rasch model we also can apply the conditional maximum likelihood (CML). It turns out that in discussing CML another model representation is preferable. We rewrite the Rasch model as

$$P_i(\theta) = \frac{\exp(\theta - b_i)}{1 + \exp(\theta - b_i)} = \frac{\xi\varepsilon_i}{1 + \xi\varepsilon_i} \tag{9.23}$$

where $\xi = \exp(\theta)$ and $\varepsilon_i = \exp(-b_i)$.

Assume that we have two items with item parameters ε_1 and ε_2. The probability that item 1 is correctly answered, given one correct response on the two-item test is equal to the probability of the score pattern (1,0), divided by the sum of the probabilities of the score patterns (1,0) and (0,1). The probability is

$$P(x_1 = 1 \mid x_1 + x_2 = 1, \xi) = \frac{\dfrac{\xi\varepsilon_1}{1 + \xi\varepsilon_1} \times \dfrac{1}{1 + \xi\varepsilon_2}}{\dfrac{\xi\varepsilon_1}{1 + \xi\varepsilon_1} \times \dfrac{1}{1 + \xi\varepsilon_2} + \dfrac{1}{1 + \xi\varepsilon_1} \times \dfrac{\xi\varepsilon_2}{1 + \xi\varepsilon_2}} = \frac{\varepsilon_1}{\varepsilon_1 + \varepsilon_2} \tag{9.24}$$

The probability in Equation 9.24 is independent of the value of the person parameter ξ. The comparison between items can be made independent of the value of the person parameters (and vice versa the comparison between person parameters can be made independent of the items). The measurements are *specific objective*.

With three items we can do the same thing as with two items. The probability that item 1 is correct given a total score $t = 1$ on the three-item test is $P(x_1 = 1, x_2 = 0, x_3 = 0 \mid t = 1) = \varepsilon_1/(\varepsilon_1 + \varepsilon_2 + \varepsilon_3)$.

The probability that item 1 is correct given a total score equal to 2 is $P(x_1 = 1, x_2 = 1, x_3 = 0 \mid t = 2) + P(x_1 = 1, x_2 = 0, x_3 = 1 \mid t = 2) = (\varepsilon_1\varepsilon_2 + \varepsilon_1\varepsilon_3)/(\varepsilon_1\varepsilon_2 + \varepsilon_1\varepsilon_3 + \varepsilon_2\varepsilon_3)$.

This result can be generalized to a test with n Rasch items. First, some notation is introduced. In the denominator of $P(x_i = 1 \mid t = 2)$, all combinations $\varepsilon_i\varepsilon_j$ appear; in the numerator only the combinations with the parameter of item i, ε_i, are entered. The sum of the products is called an elementary symmetric function. The elementary symmetric function for four items and a total score equal to 2, the elementary symmetric function of order 2, is $\gamma_2 \, (\varepsilon_1,\varepsilon_2,\varepsilon_3,\varepsilon_4) = \gamma_2(\varepsilon) = \varepsilon_1\varepsilon_2 + \varepsilon_1\varepsilon_3 + \varepsilon_1\varepsilon_4 + \varepsilon_2\varepsilon_3 + \varepsilon_2\varepsilon_4 + \varepsilon_3\varepsilon_4$.

The denominator of $P(x_i = 1 \mid t = 2)$ can be written as $\gamma_2(\varepsilon)$. For the numerator of $P(x_i = 1 \mid t = 2)$ $\gamma_1^{(1)}(\varepsilon)$, the elementary symmetric function of order 1 exclusive item 1, is needed. The numerator can be written as $\varepsilon_1\gamma_1^{(1)}(\varepsilon)$. The elementary symmetric function of order 0 is defined to be equal to one.

In the general case there are n items. The number of correct responses s_i to item i $(i = 1,...,n)$ is based on the group of N persons with $0 < t < n$ responses correct. The conditional likelihood is equal to

$$L_{\mathrm{CML}} = \mathrm{Prob}(s_1,...,s_n \mid t_1,...,t_N) = \frac{\mathrm{Prob}(s_1,...,s_n,t_1,...,t_N)}{\mathrm{Prob}(t_1,...,t_N)}$$

$$= C \frac{\displaystyle\prod_{i=1}^{n} \varepsilon_i^{s_i}}{\displaystyle\prod_{p=1}^{N} \gamma_{t_p}(\delta)} \tag{9.25}$$

where C is a factor independent of the item and person parameters. The estimation equation for the estimation of ε_i is

$$s_i = \sum_{t=1}^{n-1} N_t P(x_i = 1 \mid t) = \sum_{t=1}^{n-1} N_t \frac{\varepsilon_i \gamma_{t-1}^{(i)}(\delta)}{\gamma_t(\delta)} \tag{9.26}$$

where N_t denotes the number of persons with score t $(t = 1,...,n - 1)$. So, in CML the item score s_i is set equal to the expected item score based on the conditional probabilities correct for the various total scores. The estimation equations (Equation 9.26) for the item parameters ε_i $(i = 1,...,n)$ are solved iteratively under one constraint needed in order to fix the latent scale.

9.11 More on the estimation of item parameters

Special attention must be given to the presence of missing values when item parameters are estimated. There are two cases to consider: data can be missing by design or not. Data are missing by design when, for example, the number of items is too large to present all items to an examinee. Then subtests can be administered to different examinee groups. The subtests must have common items in order to obtain item and person parameter estimates on a common scale (see Chapter 11). All maximum likelihood methods can be generalized to incorporate values missing by design. For the MML approach, the consequence is that latent distributions for multiple groups must be defined. Data also can be missing because items are skipped. In achievement testing, skipped items sometimes are treated as incorrect responses. Another, more adequate approach to deal with skipped items with the multiple-choice format is suggested by Lord (1980). When the presence of missing values correlates with latent ability, data are not missing at random (Little and Rubin, 1987). MML estimation of item parameters is affected. This is the case when the test is speeded—that is, when some examinees do not reach the items at the end of the test, or when the time limit on the test stimulates strategic answer behavior and examinees rapid-guess on (more difficult) items (Wang and Zhang, 2006; Wise and DeMars, 2006). Then blind application of, for example, the 3PL model is inadequate.

For accurate estimation of a difficulty parameter, it is important that the group of persons that took the test has an average ability level comparable to the item difficulty. In the 2PL model and the 3PL model, discrimination parameters must be estimated. These parameters define the slopes of the ICCs. Information on the steepness of a slope is available only when the latent abilities are reasonably well spread. The Rasch model does not have a discrimination parameter. In the Rasch model, item parameter estimation can be accurate even if all persons have the same ability. This advantage is of limited value, however. If all abilities are equal, there is no way of discriminating between alternative models. The estimation of c in the 3PL model is more accurate when we have more relatively low abilities. Inaccurate estimation of the pseudo-chance-level parameter has an impact on the estimation of the discrimination parameter and the difficulty parameter as well, for the estimates of the item parameters are correlated. For known abilities the inverse of the matrix of error variances and covariances of the item parameter estimates, the information matrix of the item parameters, is given in Lord (1980).

The CML estimation procedure for the Rasch model has a clear statistical advantage above the other estimation procedures as was discussed in the previous section. For the Rasch model, software was developed for the estimation of item parameters with CML; CML estimation also can be done with a special kind of log-linear analysis

(Heinen, 1996; Kelderman, 1984). CML is, however, computationally demanding. This problem might be avoided by using MML in which the characteristics of the population distribution are estimated along with the item parameters (De Leeuw and Verhelst, 1999). There is another disadvantage of using software for the Rasch model. If the Rasch model does not fit the data very well, one could consider fitting other models and in that case other software is needed. The Rasch model can be viewed as a submodel of the 2PL model and the 3PL model. There is much to say for using the same software to compute item parameter estimates for alternative models and to compare the outcomes of these models.

Software for the analysis of item responses is commercially available, for well-known item response models like the IRT models for dichotomous data discussed here as well as for other models. Information on software can be found in books and articles that describe applications or research with the software, from software houses, and from software review sections of journals like *Applied Psychological Measurement*. Embretson and Reise (2000), who introduce many of IRT models, discuss a selection of the commercially available computer programs:

TESTFACT (Wilson, Wood, and Gibbons, 1991; www.ssicentral.com) for the full-information factor analysis of dichotomous data with the two- and three-parameter normal ogive models (with fixed values c)

BILOG (Mislevy and Bock, 1990) for the estimation of the 1PL, 2PL, and 3PL models, and BILOG-MG (Zimowski, Muraki, Mislevy, and Bock, 1996; www.ssicentral.com) for the analysis of multiple groups (BILOG is no longer available; see BILOG-MG3 of Assessment Systems Corporation)

MULTILOG (Thissen, 1991; www.ssicentral.com) and PARSCALE (Muraki and Bock, 1997; www.ssicentral.com) for dichotomous as well as polytomous items

XCALIBRE (Assessment Systems Corporation, 1996; www.assess.com) for the estimation of parameters in the 2PL and 3PL model

RUMM (Andrich, Sheridan, and Luo, 2000; www.rummlab.com) for the estimation of parameters of various Rasch models

The authors notice the fact that no final review of software is possible, because programs have been revised and will be revised continually. They also notice that alternative programs may unexpectedly produce different results although model specifications are identical.

So, more comparative studies on IRT programs and possible flaws of certain programs have to be done. It is to be hoped that this leads to improvements of the IRT software.

For WINSTEPS and information on other software packages for Rasch analyses, see www.winsteps.com.

9.12 Maximum likelihood estimation of person parameters

Once item parameters have been estimated, with CML or MML, person parameters can be estimated by maximum likelihood given the estimated item parameters (i.e., in a way similar to estimation in JML). In this section we will discuss the estimation of person parameters by ML (for bias in ML estimation, see Warm, 1989).

The person parameter in each of the models can be obtained by solving Equation 9.15 for θ. We rewrite this equation as

$$\sum_{i=1}^{n} w_i(\theta)x_i = \sum_{i=1}^{n} w_i(\theta)P_i(\theta) \tag{9.27}$$

with

$$w_i(\theta) = \frac{P_i'(\theta)}{P_i(\theta)[1 - P_i(\theta)]} \tag{9.28}$$

where $P_i'(\theta) = \partial P_i(\theta)/\partial \theta$.

In the equation a weighted total score is set equal to the weighted sum of the expected item scores. The weight in Equation 9.28 can be compared to the optimal weight in Equation 4.8 for congeneric measurements.

In the 3PL model the size of the weight $w_i(\theta)$ depends on the value of θ. For $c_i > 0$, the weight decreases as θ becomes smaller; at $\theta = -\infty$ the weight is zero. In the 3PL model, one must verify whether the solution to Equation 9.27 is a maximum. Equation 9.27 is always solved by setting θ equal to $-\infty$, but in some aberrant cases there is a maximum of the likelihood at $\theta = -\infty$. In some cases, with aberrant response patterns two maxima exist, one for $\theta = -\infty$.

For the 2PL model, optimal weight (Equation 9.28) is equal to the discrimination parameter a_i. In the Rasch model, in which the discrimination parameters are equal, all weights are equal to 1. In the Rasch model the left-hand side of Equation 9.27 reduces to the total score; the total score is sufficient for the estimation of θ (see Equation 9.17).

Several applications of IRT models presuppose the ML approach. The use of the information function as a measure of precision presupposes maximum likelihood estimation of θ. In computerized adaptive testing (Chapter 10), either the maximum likelihood estimate of θ or a Bayesian estimate, the subject of the next section, is used.

In some applications, such as the scoring of a group of examinees on the same test, there is no obligation to use maximum likelihood. Then the question arises whether to use optimal scoring weights or not. For the Rasch model, this question is easily answered. In this model, the maximum likelihood estimator is a nonlinear transformation of the unweighted total score on the test. So, total score can be computed and reported either on the observed score scale or on a transformation of this scale. Things are different with respect to the 3PL model and the 2PL model.

With the 2PL model, the optimal item weights are equal to the item discrimination parameters a_i. One might wonder how much gain in accuracy is obtained by using these weights instead of the unweighted total score. This is of some importance, for the unweighted total score has the advantage that the scoring rule needs not much explaining to the examinees. There is a second reason to be careful with using the optimal weights. The optimality of the weights is based on the assumption that the item parameters are accurately estimated and that the IRT model fits the data.

It is easy to imagine that in many situations the gain from using optimal weights is only apparent. If the discrimination parameters a_i differ moderately, the gain in accuracy when using weighted scores is limited. An additional problem can be inaccuracy of the item parameter estimates. Using "optimal" weights based on inaccurate parameter estimates might result in less accurate estimates of abilities than unit weighting. Then it is advisable to use unit weights just for statistical reasons.

In the 3PL model there is a further complication in connection with the differential weighting of items: the optimal item weight does not depend only on the item parameters, but also on the unknown person parameter. Fortunately in some applications optimal weights can be chosen that do not depend on ability level. This is the case when differentiation between ability levels is really needed only near a particular ability level, θ_0. In that case, weights $w_i(\theta_0)$ are appropriate for all values of θ (Lord, 1980, p. 170) and the problem of weighted scores is equal to the problem of weighting in the 2PL model.

Tests have been scored with unit item weights even though methods for weighting items have been proposed many times, frequently
for good reasons. An IRT analysis remains very useful even when it
has been decided to use unit weights for the combination of the item
scores. In Chapter 6 it was demonstrated that the outcome of an IRT
analysis can be used for the computation of the conditional standard
error of measurement.

9.13 Bayesian estimation of person parameters

In MML we work with a distribution of θ, with characteristics fixed
beforehand or estimated from the test data. Prior knowledge of the
distribution can be used to obtain a Bayesian estimate of θ.

With a discrete distribution of θ we can write the posterior probability distribution for θ given the responses to the items as

$$P(\theta_k \mid \mathbf{x}) = \frac{P(\mathbf{x} \mid \theta_k)g(\theta_k)}{P(\mathbf{x})} = \frac{P(\mathbf{x} \mid \theta_k)g(\theta_k)}{\sum_{h=1}^{q} P(\mathbf{x} \mid \theta_h)g(\theta_h)}, \quad k = 1,\dots, q \qquad (9.29)$$

where $g(\theta_k)$ is the prior probability of θ_k and $P(\mathbf{x}\mid\theta) = L(\mathbf{x}\mid\theta)$ is the
likelihood of the observed scores $\mathbf{x} = (x_1,\dots, x_n)$ given θ.

There are two alternative Bayesian estimators of θ, the posterior
mean or EAP estimator (*expected a posteriori estimator*) and the posterior mode. With a discrete distribution the EAP estimator is the
obvious choice. The estimator is

$$\text{EAP}(\theta) = \sum_{k=1}^{q} \theta_k P(\theta_k \mid \mathbf{x}) \qquad (9.30)$$

The posterior mean is very easy to compute: no iterations are needed
in order to obtain it. The posterior variance can be computed as a
measure of uncertainty.

9.14 Test and item information

In classical test theory the variance of measurement errors was a
relevant concept. The inverse of the error variance is an indicator of
the precision with which statements about persons can be made. In

IRT it appears advantageous to start with the precision of measurements. The two central concepts are *test information* and *item information*.

The test information at a level of latent ability θ, $I(\theta)$, gives, under some conditions, the precision with which θ can be estimated at this ability level. The conditions are as follows:

1. We have chosen the adequate model.
2. The item parameters are accurately estimated.
3. We use maximum likelihood estimation—that is, we use optimal weights for the estimation of θ; with other estimation methods (e.g., with number right scoring) we speak of the information of a scoring formula (Birnbaum, 1968); the latter information cannot exceed the test information.
4. The test is not too short.

The test information has a very convenient property. It is the sum of the item informations (Birnbaum, 1968). So, the contribution of each item to the accuracy of a test may be considered apart from the contributions of other items. The test information is

$$I(\theta) = \frac{\left[\sum_i w_i(\theta) P_i'(\theta) \right]^2}{\sum_i w_i(\theta)^2 P_i(\theta)[1 - P_i(\theta)]} = \sum_i I_i(\theta) \qquad (9.31)$$

where $w_i(\theta)$ is the optimal item weight from Equation 9.28, $P_i'(\theta)$ is the derivative of $P_i(\theta)$ with respect to θ, and $I_i(\theta)$ is the item information of item i,

$$I_i(\theta) = \frac{P_i'(\theta)^2}{P_i(\theta)[1 - P_i(\theta)]} \qquad (9.32)$$

The item information is equal to the square of the slope of the ICC at θ divided by the local error variance, which is the item variance given θ. For polytomous items, results are given in Exhibit 9.3.

Exhibit 9.3   Optimal weights and information of polytomous items

Optimal weight (Equation 9.28) and item information (Equation 9.32) cannot directly be generalized to the case of polytomous items. With polytomous items, each option has an optimal option weight. Let $P_{ik}(\theta)$ be the option characteristic curve. Then the optimal weight associated with option k is

$$w_{ik}(\theta) = \frac{\partial \ln P_{ik}(\theta)}{\partial \theta} = \frac{P'_{ik}(\theta)}{P_{ik}(\theta)}$$

The sum of the weights of the chosen options equals zero at the maximum likelihood estimate. The option weights function in quite a different way than the weights for dichotomous items. The reason for the difference is that the weight for a dichotomous item is an item weight: the options *correct* and *incorrect* are not weighted separately, as the score for *incorrect* is set equal to zero. The item weight is the difference between the option weight for *correct* and the option weight for *incorrect*.

For the two models in Equation 9.6 and Equation 9.7, the score weight of category k can be written as k plus a factor independent of k. The option score k can be written as

$$k = w_{ik}(\theta) - w_{i0}(\theta)$$

While k does not depend on the item and option parameters, the polytomous Rasch models have a sufficient statistic for the estimation of θ—the sum of the category numbers of the options chosen.

The item information, the sum of the option informations, is

$$I_i(\theta) = \sum_{k=0}^{m} \frac{P'^2_{ik}(\theta)}{P_{ik}(\theta)}$$

Samejima (1969) proved that under the graded response model the item information increases if categories are split into more categories.

It is easily demonstrated that the item information of a dichotomous item, as given by Equation 9.32, is a special case of the item information for polytomous items.

In the 3PL model, the item information is

$$I_i(\theta) = a_i^2 p_i(\theta)[1 - p_i(\theta)]\left(1 - \frac{c_i}{P_i(\theta)}\right) \qquad (9.33)$$

where $p_i(\theta)$ equals the probability of a correct response to the item if c_i would have been equal to 0. In the 2PL model, the item information equals a_i^2 times the item variance given θ, and in the Rasch model the item information equals the item variance given θ, the error variance on the true-score scale.

In Figure 9.8 information functions are displayed for three items. The information of the item with $c = 0.0$ and $a = 2.0$ exceeds the information of the item with the lower discrimination parameter for a large range of the latent ability. With increasing a, the information at $\theta = b$ increases, while the information at more distant abilities decreases. A perfect Guttman item discriminates at only one point—it discriminates between persons with θ smaller than b and persons with θ larger than b.

The information of the item with $c = 0.25$ and $a = 2.0$ is lower than the information of the item with the same value of a and $c = 0.0$. The reduction is lowest at high values of θ; for low values of θ the reduction is large due to guessing. From the figure, we can infer that the highest

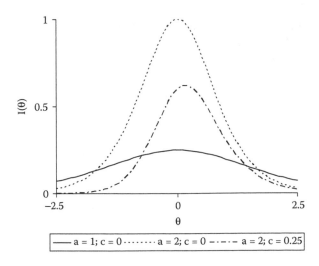

Figure 9.8 Information functions for three items ($b = 0$).

information is obtained at $\theta = b$, unless c is larger than 0. When c exceeds 0, the highest information value is obtained for a value of θ somewhat higher than b. Birnbaum (1968) gives the relationship between c and the value of θ at which the highest information is obtained.

The value of the item information and, consequently, the value of the test information, depend on the choice of the latent scale. In the 2PL model and 3PL model a linear scale transformation is allowed. We can multiply all θ and b with 2 and divide all a by 2. Then the item information and test information decrease by a factor 4 (see Equation 9.33). And, when nonlinear scale transformations are also considered—for example, a transformation to the true-score scale—the form of the information function can change dramatically.

Information is not an invariant item property. The ratio of the item informations of two items is, however, invariant:

$$\frac{I_i(\theta^*)}{I_j(\theta^*)} = \frac{I_i(\theta)}{I_j(\theta)}$$

for all monotone transformations θ^* of θ.

The *relative efficiency* of two tests, the ratio of the test informations of two tests, remains unchanged with a change of scale. This means that the comparison of the accuracy of two tests does not depend on the chosen latent scale.

The estimated value of $I(\theta)$ can be used (asymptotically) for the construction of a confidence interval for θ. The variance of $\hat{\theta}$ equals the inverse of the test information, $1/I(\theta)$, assuming accurate item parameter estimates (otherwise the error variance is larger; see, e.g., De Gruijter, 1988). Under the assumption that $\hat{\theta}$ is normally distributed, the approximate 95% confidence interval is

$$\hat{\theta} - 1.96/\sqrt{I(\hat{\theta})} < \theta < \hat{\theta} + 1.96/\sqrt{I(\hat{\theta})} \qquad (9.34)$$

With this confidence interval we might err in case a population of abilities is involved. Then we better use the EAP estimator instead of the ML estimator. With EAP we can also compute the posterior variance of θ. This variance is smaller than the inverse of the test information. These results are comparable to those discussed in connection with the application of the Kelley formula within the context of classical test theory. Also, test reliability is of foremost importance within the context of IRT. A test is useful for differentiating between persons in as far as the error variation is relatively small compared to the true variation in θ.

9.15 Model-data fit

An important question is whether the chosen IRT model fits the data. If the model does not fit, we have to find a less restrictive model that fits the data or we have to drop nonfitting items. The investigation of model fit has two aspects:

1. A statistical test of model fit
2. An analysis of residuals, in order to get an idea of the seriousness of the model violations

With a small number of data, it is difficult to reject a particular model. With a large number of data, a model is easily rejected even if the deviations of the data from the model are small for all practical purposes. For this reason, both statistical testing and the analysis of residuals are necessary in the study of model fit.

In the models we discussed so far, we have one general model assumption: the assumption of unidimensionality or local independence. For each model we also have assumptions regarding the specific shape of the item response curves. The assumption of unidimensionality can be verified in several ways.

A factor analysis might reveal whether the data are unidimensional. A factor analysis of phi coefficients—ordinary product moment correlations—is not suitable when the item discriminations are high. The analysis of phi coefficients would produce spurious factors. A multidimensional IRT analysis, which is a nonlinear factor analysis, is called for. A factor analysis of tetrachoric correlations—correlations of bivariate normal distributions assumed to underlie the responses to pairs of dichotomous items—might be appropriate when the latent distribution is normal and there is no guessing.

We can examine the item variances and covariances after the elimination of the model effects. A positive correlation between residuals indicates violation of model assumptions. Stout (1987) presented a nonparametric test for unidimensionality based on a split of the test into two subtests (see also Roussos, Stout, and Marden,1998). The research by Stout and his coworkers resulted in programs like (POLY-)DIMTEST and DIMTEST. An overview of the work can be found in Stout (2002; for software like DETECT and DIMTEST, see www.assess.com). A comparison of several nonparametric approaches to dimensionality checking is presented by Bouwmeester and Sijtsma (2004).

Specific model assumptions can be verified in various ways. Varying values of item-test correlations might indicate different discrimination parameters. The study of item-test regressions can reveal whether guessing is likely to play a role. We also can do an analysis of residuals by item, after having completed an IRT analysis. First we group the estimated θ's in a number of sufficiently large groups. Next in each group the (standardized) difference between the observed proportion correct p and the model probability is computed. Yen's (1981) statistic Q_1 is based on the squared differences. The statistic she proposed is a Pearson chi-square. For item i the statistic can be written as

$$Q_{1i} = \sum_{l=1}^{m} \frac{N_l[p_{il} - P_i(\theta_l)]^2}{P_i(\theta_l)[1 - P_i(\theta_l)]} = \sum_{l=1}^{m} z_{il}^2 \qquad (9.35)$$

where $P_i(\theta_l)$ stands for the average probability correct in ability group l and N_l is the size of ability group l. Q_{i1} is approximately distributed as a chi-square with $m - k$ degrees of freedom, where m equals the number of score groups and k is the number of item parameters. A likelihood ratio statistic has been proposed by McKinley and Mills (1985). This statistic can be written as

$$G_i^2 = -2\ln\frac{L_{i1}}{L_{i0}} = 2\sum_{l=1}^{m}\left(N_{il} \ln \frac{p_{il}}{P_i(\theta_l)} + (N_l - N_{il})\ln \frac{1 - p_{il}}{1 - P_i(\theta_l)} \right) \qquad (9.36)$$

where L_1 is the likelihood given the model, L_0 the likelihood without model restrictions, and N_{il} the number of correct responses to item i in group l. This statistic is also approximately chi-square distributed. A problem with the Q_1 and G statistics is that the grouping of persons is based on estimated values θ. In a simulation study on new tests of fit and the G statistic from Equation 9.36 by Glas and Falcón (2003), it was demonstrated that this G statistic has an inflated type I error rate.

Figure 9.9 shows the fit of an item of the verbal analogies test, a subtest of a Dutch intelligence test, the NDT or Netherlands Differentiation Test, using the likelihood ratio fit test (Equation 9.36). The

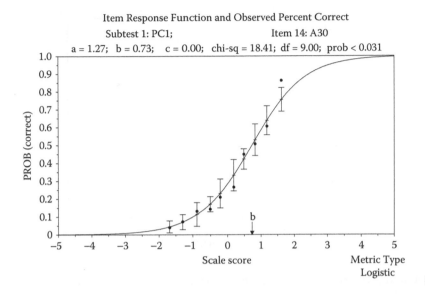

Figure 9.9 Item fit for an item of a Verbal Analogies Test of the NDT, the Netherlands Differentiation Test (BILOG output).

figure includes the item response curve and the empirical regression of the item.

Other chi-square statistics, on the item and the test levels, have been proposed for the Rasch model, based on the CML approach (Andersen, 1973; Kelderman, 1984; Molenaar, 1983).

In MML an overall test to compare nested models seems to be possible (Reise, Widaman, and Pugh, 1993).

The residuals used for the computation of item fit statistics might also be plotted. Graphical model control might add useful information on the cause of item misfit. For other checks on model fit, graphical methods are useful. An example is given in Figure 9.10 where Rasch item parameters have been estimated in a group with high scores (H) and a group with low scores (L). The item parameter estimates are not invariant. In the low scoring group, the range of values of the item parameter estimates is smaller. This is due to the fact that guessing played a role in the (simulated) data. A good overview of graphical methods is given by Hambleton, Swaminathan, and Rogers (1991).

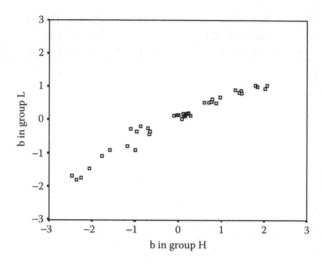

Figure 9.10 Rasch item parameter estimates, a low-scoring group versus a high-scoring group.

9.16 Appendix: Maximum likelihood estimation of θ in the Rasch model

In this appendix we give an example of the way maximum likelihood parameter estimation proceeds. For simplicity, we chose the estimation of the person parameter in the Rasch model.

The likelihood of a score pattern $\mathbf{x} = (x_1,..., x_n)$ in the Rasch model can be written as

$$L(\mathbf{x}\,|\,\theta) = \prod_{i=1}^{n} \frac{\exp(\theta-b_i)^{x_i}}{1+\exp(\theta-b_i)} = \exp(\theta)^t \prod_{i=1}^{n} \exp(-b_i)^{x_i} \prod_{i=1}^{n} [1+\exp(\theta-b_i)]^{-1}$$

(9.37)

where t is the total score. We want to find the value of θ that maximizes Equation 9.37. Maximizing the likelihood is equivalent to maximizing the logarithm of the likelihood. So, instead of maximizing Equation 9.37, we maximize its logarithm:

$$\ln L(\mathbf{x}\,|\,\theta) = t\theta - \sum_{i=1}^{n} x_i b_i - \sum_{i=1}^{n} \ln[1+\exp(\theta-b_i)]$$

(9.38)

where ln is the natural logarithm. When ln $L(\mathbf{x}|\theta)$ has obtained its maximum as a function of θ, the derivative of ln $L(\mathbf{x}|\theta)$ with respect to θ is equal to 0. So we can find the ML estimate of θ by differentiating Equation 9.38 with respect to θ and setting the result equal to 0 (we must check whether we have obtained a maximum of the function and not a minimum).

Differentiating Equation 9.38 with respect to θ and setting the result equal to 0 gives the equation

$$g(\theta) = t - \sum_{i=1}^{n} \frac{\exp(\theta - b_i)}{1 + \exp(\theta - b_i)} = t - \sum_{i=1}^{n} P_i(\theta) = 0 \qquad (9.39)$$

This equation is identical to Equation 9.17.

In Figure 9.11, the likelihood, the logarithm of the likelihood, and the derivative of ln $L(\mathbf{x}|\theta)$ are displayed for a simple example. In Figure 9.11a, we can see that two response patterns with identical total scores have very different values for the likelihood, but also that for both response patterns the maximum is obtained for the same value of θ (0.721). In Figure 9.11b, ln $L(\mathbf{x}|\theta)$ is given. From Figure 9.11a and Figure 9.11b, it is clear that the value of θ that maximizes $L(\mathbf{x}|\theta)$ also maximizes ln $L(\mathbf{x}|\theta)$. For this value of θ, $g(\theta)$ in Figure 9.11c equals 0.0. In Figure 9.11c, we see that the slope of $g(\theta)$ is negative. This shows that we obtained a maximum instead of a minimum of ln $L(\mathbf{x}|\theta)$.

We find θ by solving Equation 9.39. We have a maximum in Equation 9.39 if $g(\theta)$ decreases with increasing θ in the neighborhood of $g(\theta)$ = 0.0. In other words, the derivative of $g(\theta)$ with respect to θ, the second-order derivative of ln L, must be negative. The derivative of $g(\theta)$, $g'(\theta)$, can be used in the estimation procedure. This is done in the Newton–Raphson procedure.

Let us demonstrate the estimation procedure. We approximate the function $g(\theta)$ at the value of the estimated θ in iteration k, θ^k, by a straight line. This line goes through the point $(\theta^k, g(\theta^k))$ and has a slope equal to $g'(\theta^k)$. The equation of this line is

$$g(\theta) = g(\theta^k) + g'(\theta^k)(\theta - \theta^k) \qquad (9.40)$$

Setting $g(\theta)$ equal to 0 gives us the next estimate of θ:

$$\theta^{k+1} = \theta^k - g(\theta^k)/g'(\theta^k) \qquad (9.41)$$

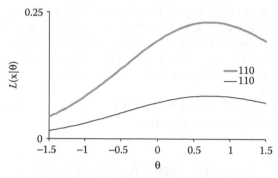

(a) $L(\mathbf{x}|\theta)$; The likelihoods of two of the three response
patterns with a total score equal to 2 are given

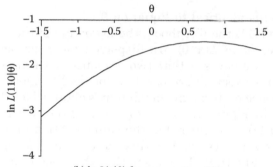

(b) $\ln L(\mathbf{x}|\theta)$ for score pattern 110

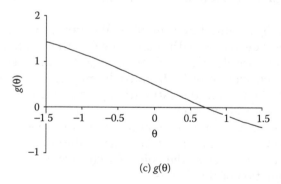

(c) $g(\theta)$

Figure 9.11 The likelihood $L(\mathbf{x}|\theta)$, $\ln L(\mathbf{x}|\theta)$, and the derivative of $\ln L(\mathbf{x}|\theta)$
with respect to θ for a test with $b_1 = -0.5$, $b_2 = 0.0$, $b_3 = 0.5$, and a total score
equal to 2.

In the Newton–Raphson method, we need $g'(\theta)$, the derivative of $g(\theta)$ in Equation 9.39:

$$g'(\theta) = -\sum_{i=1}^{n} P_i(\theta)[1 - P_i(\theta)] \qquad (9.42)$$

The derivative of $g(\theta)$ is equal to minus the test information. This relation between the second-order derivative of the log likelihood, $g'(\theta)$, and the test information does not hold for the 3PL model. With this model, the test information only equals minus the *expected* value of $g'(\theta)$ (Kendall and Stuart, 1961, pp. 8–9). In the 3PL model, the test information is easier to compute than the second-order derivative $g'(\theta)$, so with this model we replace the second derivative in Equation 9.40 by minus the test information (using Fisher's method of scoring instead of the Newton–Raphson procedure).

The iterative procedure for the estimation of θ in the Rasch model is as follows:

A. We compute a starting value for θ, θ^0.
B. We compute a new value θ^1 by application of Equation 9.41:

$$\theta^1 = \theta^0 - \frac{\left(t - \sum_{i=1}^{n} P_i(\theta^0) \right)}{\left(-\sum_{i=1}^{n} P_i(\theta^0)[1 - P_i(\theta^0)] \right)} \qquad (9.43)$$

C. We compute $|\theta^1 - \theta^0|$, the absolute value of the difference between the two consecutive estimates of θ.
D. If the value obtained in step C is below a chosen threshold value ε, we stop. The obtained value θ^1 is our final ML estimate. If the difference exceeds the threshold, we replace θ^0 by θ^1 and repeat steps B and C. This process is repeated until we reach convergence (by the way, with an inadequate starting value, the procedure may fail to converge).

Let us give a numerical example of the method. We have three items, with $b_1 = -0.5$, $b_2 = 0.0$; and $b_3 = 0.5$. The total score t equals 2. As the starting value, we choose $\theta = 0.0$. The final estimate of θ is 0.721, obtained in the second iteration. The data of the iteration

Table 9.1 Iteration history.

Iteration	θ^k	$g(\theta^k)$	$g'(\theta^k)$	θ^{k+1}
$k = 0$	0.0	0.5	−0.72001	0.69444
$k = 1$	0.69444	0.01706	−0.64820	0.72075
$k = 2$	0.72075	0.00007	−0.64304	0.72086

process are given in Table 9.1. You might want to verify these figures yourself using a spreadsheet.

Exercises

9.1 Compute the probability of a correct response for a Rasch item with item parameter equal to 0.0 and person parameter $\theta = -2.0$ (0.5) 2.0.

9.2 We have the responses of two homogeneous groups of persons on two items. The response probabilities are $P_1(\theta_1) = 0.3775$, $P_1(\theta_2) = 0.6225$, $P_2(\theta_1) = 0.4378$, and $P_2(\theta_2) = 0.7112$. Estimate the person parameters θ_1 and θ_2 on the basis of the response probabilities for the first item, assuming that the Rasch model fits. Use the response probabilities for the second item for verifying whether the Rasch model really fits.

9.3 Given is a test with three Rasch items. The item parameters are $b_1 = -0.5$, $b_2 = 0.0$, and $b_3 = 0.5$. A person has answered items 1 and 2 correctly, and item 3 incorrectly. Compute the likelihood for $\theta = -1.0, -0.5, 0.0, 0.5, 1.0$.

a. Consider the four intervals defined by the five values of θ for which the likelihood has been computed. In which interval lies the maximum likelihood estimate of θ?

b. Assume that we have a population distribution with five latent classes: $\theta_1 = -1.0$, $\theta_2 = -0.5$, $\theta_3 = 0.0$, $\theta_4 = 0.5$, $\theta_5 = 1.0$. Also assume that these latent classes have the same relative frequencies: $g(\theta_k) = 0.2$ for $k = 1,..., 5$. Compute the EAP estimate of θ.

9.4 We have three items with item parameters:
$b_1 = 0.5$, $a_1 = 1.0$, $c_1 = 0.0$
$b_2 = 0.5$, $a_2 = 2.0$, $c_2 = 0.0$
$b_3 = 0.5$, $a_3 = 2.0$, $c_3 = 0.25$
Compute the item informations at $\theta = 0.0$.

9.5 We have a discrete distribution of θ with values -1, -0.5, 0.0, 0.5, and 1. The following is known:

Value θ	Frequency $f(\theta)$	$I(\theta)$
−1.0	0.1	7.864
−0.5	0.2	9.400
0.0	0.4	10.000
0.5	0.2	9.400
1.0	0.1	7.864

Compute the reliability of the test when maximum likelihood is used for the estimation of θ.

CHAPTER 10

Applications of Item Response Theory

10.1 Introduction

Item response theory (IRT) can be used to analyze a test and to perform an item analysis. In Section 10.2, item analysis with IRT will be discussed. There is more to IRT. The development of IRT has opened new ways for test applications and research with tests. This is true especially for the parametric IRT models. Nonparametric models may be less restrictive than parametric models, but they are also less informative.

An important application of IRT is test equating, or bringing test scores to the same scale. IRT equating has greatly extended the possibilities of equating. Large-scale testing programs also profit from developments in IRT. Different persons may get different, partially overlapping tests. Test results from all persons and all items can be brought to the same scale using IRT. We will devote the next chapter to equating with and without IRT.

Item banking is an important tool with IRT-based testing. Assume that a new test is administered that contains old items, items with known item parameters, and also a number of new items. After the test administration we can estimate the item parameters of the new items on the common latent scale as well. By repeatedly applying this procedure, we are building a large pool of items with known item parameters. As long as the item characteristics do not change due to, for example, educational change (the phenomenon of item drift; Donoghue and Isham, 1998), we have an *item bank* from which we may choose items with known item parameters at will.

One application of item banks is with test construction. In the application of standard test-development techniques of classical test theory to the construction of tests, items are selected on the basis of two statistical characteristics: item difficulty and item discrimination. What IRT has to offer for test development in general, and item selection in particular, is described in Section 10.3.

IRT enables us also to investigate item bias (Section 10.4) and inappropriate response patterns (Section 10.5) in a better way than classical test theory.

Another important application of IRT is in so-called computerized adaptive testing. Because under ML scoring items can be viewed as the building blocks of tests, an item bank enables us to administer computerized adaptive tests (CAT), the subject of Section 10.6.

The use of IRT in the measurement of change is discussed in Section 10.7. Finally, IRT makes it possible to tackle various problems by doing simulation studies. An example, discussed in Exhibit 10.1, is the determination of the optimal number of options in multiple-choice items. In Section 10.8, some concluding remarks on IRT applications are made. Software for some of the applications mentioned in this chapter can be obtained from, for example, www.assess.com.

Exhibit 10.1   On the optimal number of options in multiple-choice items

Items with four answer options are more accurate than items with three or two options, ceteris paribus—the effect of guessing is smaller with these items. On the other hand, more two-option items can be answered in the same testing time than four-option items. So, it is a sensible question to investigate whether one should administer a test with, say, two-option items rather than a test with four-option items. There have been studies on this topic from both a theoretical as well as an empirical perspective. Let us follow Lord's (1977) line of reasoning.

First, we must decide how many items with a particular number of options can be administered within a given testing time. Lord assumed that the reading time depends on the number of options alone (i.e., the number of items times the number of options is fixed for a given testing time). This assumption is testable for a particular application.

Lord also assumed that item characteristics do not change with a change of the number of choices, except the value of the pseudo-chance-level parameter. The pseudo-chance-level parameter c is set equal to the inverse of the number of options, although this decision is not supported by the outcome of IRT research.

Lord showed that three-option items provide the most information at the midrange of the scale score, whereas the two-option item works best at the upper range. When pass–fail decisions must be made, tests with two or three options are optimal given optimal item difficulties and the validity of the assumptions mentioned above.

The bulk of the research on the optimal-number-of-options problem is done from an empirical perspective. One such study is the study by Rogers and Harley (1999). Their overall conclusion is that tests consisting of three-option items are at least equivalent to tests composed of four-option items in terms of internal consistency. Haladyna (1999) strongly recommends three-option items instead of four- or five-option items. His recommendations, however, fell on deaf men's ears. The bulk of test constructors and item writers stick to four or five options with multiple-choice items.

10.2 Item analysis and test construction

IRT can be used to investigate the dimensionality of a test and to screen the items of the test. Some remarks with respect to item analysis have been made in Chapter 6. In this section, we proceed by giving two examples of IRT analyses of tests.

IRT modeling is frequently used with achievement testing. Our first example is about the IRT analysis of measurement instrument from another domain. The instrument is a personality questionnaire, the Multidimensional Personality Questionnaire (Tellegen, 1982), and the IRT analysis done by Reise and Waller (1990).

Reise and Waller did several analyses of the scales of the MPQ. They did, for example, a factor analysis on the tetrachoric correlations. They concluded that the responses on each scale could be accounted for by one dominant dimension. The responses of two samples of 1000 persons were analyzed with the one- and two-parameter logistic models. The overall fit of the two-parameter model was adequate, although some items did not fit well. Reise and Waller concluded that the IRT analysis gave more information on the psychometric properties of the scales and the items than would have been possible with an analysis based on the classical test model.

The second example is about the application of IRT in the construction of a test to measure early mathematical competence. Van der Rijt, Van Luit, and Pennings (1999) describe the construction of two versions or scales of the Utrecht Early Mathematical Competence Scales. The scales were developed in order to assess the developmental level of early mathematical competence in children ages 4 to 7 years. First, items were written for eight aspects of numerical and nonnumerical knowledge of quantity. This resulted in a pool of 120 items, from which two 40-item scales had to be constructed.

The total test of 120 was too large to be presented to the children participating in the investigation. So, several test booklets were constructed. Common items would make it possible to obtain item parameters on a common scale. First, item difficulties were computed and a factor analysis was done on the scores of the eight aspects. The results of the factor analysis suggested that there is one underlying dimension of mathematical competence.

The data were analyzed with the Rasch model. This model was rejected. Next, an analysis was done with the two-parameter model in which the values of the discrimination parameters a were fixed. With fixed discrimination parameters, the item difficulty parameters could be obtained with a CML analysis (Verhelst and Glas, 1995). Some items did not fit well. Among the nonfitting items were items that were frequently guessed correctly. The nonfitting items were eliminated from the item pool, and the two 40-item scales were composed of items from the pool.

10.3   Test construction and test development

In this section, it is assumed that a large set or pool of items is available for test construction (Vale, 2006), and that accurate item parameter estimates of these items were obtained. This could have been achieved by an analysis of the responses of examinees to different test forms with common items.

When good estimates of the item parameters are available, we can compute the item information. The item information has the additivity property under ML scoring of persons. The sum of the item informations produces the test information. This entails that we can compute test characteristics beforehand from the characteristics of the items that we choose for the test. So, we can construct the shortest test that at a certain ability level has a test information exceeding a specified minimum value. To simplify matters, we assume that there are no additional restrictions on the test composition.

Suppose that we want to construct a test that has an error variance of maximally the value d at ability level θ_0. Then the test information at that point, $I(\theta_0)$, should have at least the value $1/d$. We construct the shortest test as follows. We take as first item in the test the item with the highest value of the item information at θ_0. As second item we choose the item that has the highest item information among the remaining items. We go on with selecting items until the sum of the item informations at θ_0 is at least equal to $1/d$. Unfortunately, parameter estimates are not free from error. So, when selecting on item information, we might err a bit, and the computed test information might be somewhat higher than the true test information. The procedure is simple. The largest problem is the determination of the value d. When an old test is available, we could use the value d of that test.

Another possibility is that we want to specify a minimum value of the test information for two (cf. Exhibit 6.2) or more values of θ. With this conceptually simple extension, the solution becomes quite difficult.

Let us take the case where we want to discriminate for a range of θ values, and where a minimum value of $I(\theta)$ must be specified for an interval of θ. In this case, we replace the interval by some well-chosen values of θ. Next, for each of these values θ, a minimum value for the test information is specified. Now, we need to find the shortest test for which at each of these values θ the test information is at least as large as the minimum specified. This problem can be defined as an optimization problem, to be solved with linear or integer programming techniques (Theunissen, 1985; Van der Linden and Boekkooi-Timminga, 1989).

We formulate the problem as follows. Determine the minimum level for the information at θ_k, $I(\theta_k)$ $(k = 1,..., m)$. Minimize test length:

$$n = \sum_{i=1}^{N} x_i \qquad (10.1)$$

where N is the number of items in the item pool and $x_i = 1$ if the item is included in the test and 0 otherwise, subject to the constraints

$$\sum_{i}^{N} x_i I_i(\theta_k) \geq I(\theta_k) \quad (k = 1,...,m) \qquad (10.2)$$

Also, in this case, the specification of the target value of the test information might appear the most difficult part of the problem. In Chapter 6 we referred to an investigation by Cronbach and Warrington (1952), one of the first IRT-based simulation studies. They demonstrated that in most cases the highest average test accuracy is obtained with items of a similar difficulty, at the cost of a loss in accuracy at the higher and lower abilities. This is demonstrated in Figure 10.1. In Figure 10.1, the test information of two five-item tests with moderately discriminating items (all a equal to 1) are contrasted. One test is a peaked test, a test in which all b parameters are equal. In the second test, the difficulty parameters are spread, with $b = -1.5, -0.5, 0.0, 0.5$, and 1.5. The first test is optimal in a large interval around $b = 0$.

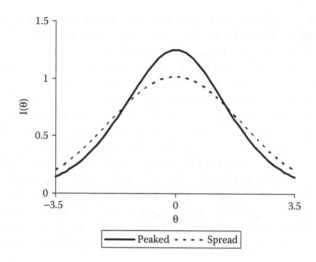

Figure 10.1 The test information of a peaked and a spread test.

The study by Cronbach and Warrington suggests that we might set the target information lower for the relatively high and low abilities.

An interesting example of the possibilities of optimal test construction is presented by Van der Linden and Reese (1998). They demonstrated the possibilities of test construction with extra constraints on, for example, subject matter coverage in the context of computerized adaptive testing using the "shadow test" approach. In this approach, at each selection step a whole test is generated that satisfies the constraints. It is quite possible that the item bank is not large enough to produce a solution to the problem under the given constraints—the problem is infeasible. There are several approaches to find the causes of the infeasibility and to force a solution nonetheless (Huitzing, Veldkamp, and Verschoor, 2005). Belov and Armstrong (2005) proposed to assemble tests using a random search approach; this approach also provides information on the most restrictive constraints.

10.4 Item bias or DIF

A test and the use of test scores are meant for a well-defined population or several distinct populations. Test takers from a population can generally be classified into several subgroups, according to, for example, gender or ethnicity, and inferences from the test scores should be

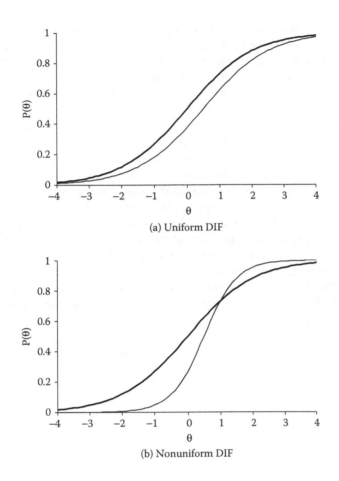

(a) Uniform DIF

(b) Nonuniform DIF

Figure 10.2 Two examples of differential item functioning (DIF); the item characteristic curves (ICCs) of one item in two different subgroups a and b.

equally valid for members of all subgroups. Bias is defined by Cole and Moss as *differential validity of a given interpretation of a test score for any definable, relevant subgroup of test takers* (1989, p. 205).

Bias can be studied in various ways. A possibility that frequently is considered is that a test is potentially biased because of the presence of biased items. So, bias can be studied by investigating the internal structure of a test.

In terms of IRT, we speak of item bias if the probability of a correct response to the item *given latent ability* differs between relevant subgroups. Figure 10.2a and Figure 10.2b illustrate item bias. In

Figure 10.2a, we have uniform item bias. One subgroup has a disadvantage at each level of θ—the ICC for this subgroup lies below the ICC for the other subgroup at each level of θ. Figure 10.2b illustrates the second possibility of item bias. In this figure, the ICCs cross. At the left of their intersection one subgroup has the advantage, at the right the other subgroup has the advantage.

To date, the more neutral term *differential item functioning* (DIF) is used in connection with group differences. In the 1999 *Standards* (APA et al., 1999), DIF is defined as a statistical property of a test item in which different groups of test takers who have the same total test score have different average item scores, or in some cases, different rates of choosing various item options (p. 175). Technically, and in IRT specifically, DIF is a special kind of violation of the unidimensionality assumption. A single latent trait (i.e., the value of θ), is not sufficient to predict the probability of a correct response; in addition, latent ability group membership is necessary in order to correctly predict the probability. This is most easily seen when in the Rasch model the biased item is shifted f units to the right for one subgroup, say subgroup 2:

$$P_i(\theta_{p(g)}) = \frac{\exp(\theta_p - b_{i(g)})}{1 + \exp(\theta_p - b_{i(g)})}$$

where $b_{i(2)} = b_{i(1)} + f$.

An alternative way of writing the ICCs for both subgroups is as follows:

$$P_i(\theta_{1p(g)}, \theta_{2p(g)}) = \frac{\exp(\theta_{1p} + \theta_{2(g)} - b_i)}{1 + \exp(\theta_{1p} + \theta_{2(g)} - b_i)}$$

where $\theta_{2(1)} = 0.0$, $\theta_{2(2)} = -f$.

If DIF is observed, it is important to identify the causes of the effect. Is the item less familiar in the focal group, the group we are interested in, than in the reference group? If so, is the differential familiarity unrelated to the ability of interest? If the difference between the performance of two groups is really due to irrelevant factors, we can conclude that the item is biased, which, of course, is undesirable. An example of research into the causes of DIF is the study by Allalouf, Hambleton, and Sireci (1999). These investigators addressed the causes of DIF in a test translated from Hebrew into

Russian. Several sources of DIF were found: changes in word difficulty, changes in item format, differences in cultural relevance, and changes in context.

Many methods for the detection of DIF have been proposed. Let us review some of them in an IRT context. Rudner, Getson, and Knight (1980) proposed to look at the size of the deviation between the ICCs in both subgroups over the relevant range of θ. In several proposals for measuring item bias, the differences between the ICCs are counted only for values of θ found in the study (Shepard, Camilli, and Williams, 1985).

In their study on item bias, Linn and Harnisch (1981) computed the item parameters on the basis of the results of all subjects in the study. For each person, they could compute the probability of a correct response given the estimate of ability. They compared the average model proportion correct for a range of values θ to the observed proportion correct in the target group. They also proposed to compute standardized differences. In this respect, their approach is similar to Yen's approach to item fit. Linn and Harnisch did not propose to use a χ^2 statistic, however. Thissen, Steinberg, and Wainer (1988) proposed a likelihood ratio statistic for an item suspected to be biased. First, they proposed to estimate the item parameters for both groups simultaneously. The resulting likelihood for this model, M_0, is L_0. Next, the item parameters can be estimated again, but the item parameters of the item under investigation are allowed to take on different values in the two groups. The likelihood for this alternative model, M_1, is L_1. The likelihood ratio statistic equals

$$LR = -2\ln\frac{L_0}{L_1} \tag{10.3}$$

LR is approximately χ^2 distributed under the null hypothesis (equal item parameters in different groups). The degrees of freedom are equal to the number of parameters set equal in both groups in model M_0 but are allowed to vary between groups in model M_1. More information on this approach is given by Kim and Cohen (1998), who use the LR test in connection with the graded response model. A factor analytic approach is discussed by Finch (2005).

Some methods for the detection of DIF are not based on the IRT approach. Two of these methods deserve our attention: *STD P-DIF* (see Exhibit 10.2), and the Mantel–Haenszel measure and chi-square statistic (see Exhibit 10.3).

Exhibit 10.2   A simple index for DIF: *STD P-DIF*

We administered a test and for one item computed the item-test regression, both for the reference group, group R, as for the focal group, group F. The latter group is the group of interest. We obtained the values in the table below for the item-test regressions (proportions correct given total score k) and score frequencies n.

Total Score	p_{Rk}	n_{Rk}	p_{Fk}	n_{Fk}
0	0.0	0	0.0	0
1	0.3000	10	0.2500	4
2	0.4000	30	0.3333	3
3	0.4588	85	0.4286	7
4	0.4818	110	0.4667	15
5	0.5133	150	0.4444	9
6	0.7143	140	0.6667	12
7	0.8538	130	0.8125	16
8	0.8800	100	0.8182	22
9	0.9556	45	0.9167	12
Total	0.6575	800	0.6600	100

$$STD\ P-DIF = \sum_k n_{Fk}(p_{Fk} - p_{Rk})/\sum_k n_{Fk} = p_F - \sum_k n_{Fk}p_{Rk}/n_F = -0.0452$$

We see from the table that four persons from the focal group have a total score equal to 1, and that one of them (a proportion equal to 0.25) has the item correct. In the reference group, the proportion correct given a total score equal to 1 is 0.3.

With *STD P-DIF* (shorthand for "standardized P-difference"), we compute the weighted mean difference between the proportions correct for the focal group and those for the reference group. As weights we use the proportions of persons from the focal group with the respective total scores. Index *STD P-DIF* has been proposed as a DIF index by Dorans and Kulick (1986; see also Dorans and Holland, 1993). *STD P-DIF* can be used to detect uniform bias. In the example above, we obtained the index for a very short test; this has been done in order to keep the computational burden small. The value of the index is –0.0452. The item is more difficult for the focal group. The proportions correct in the total groups are 0.6575 for the reference group and 0.66 for the focal group. So, in this example, the focal group has a higher overall achievement on the item, but nevertheless the item seems to be biased against this group. Actually, the effect is small. Even with large groups only standardized differences larger than 0.05 and smaller than −0.05 are considered for further inspection.

Total score has been used in the computations for the index as an indicator of ability for members of both groups. Notice too that the item that possibly displays DIF is included in the computation of the total score.

Total score is sufficient for the estimation of ability under the assumptions of the Rasch model. More importantly, in the Rasch model, the item proportion correct given total score does not depend on characteristics of the latent trait distribution.

Exhibit 10.3   The Mantel–Haenszel Procedure

At score level k, the probability of a correct response to an item is p_{1rk} in the reference group and p_{1fk} in the focal group, the group we are interested in. In the Rasch model, the odds in the reference group, $p_{1rk}/(1 - p_{1rk}) = p_{1rk}/p_{0rk}$, is equal to the odds in the focal group, p_{1fk}/p_{0fk}. So, the odds ratio is equal to 1. The Mantel–Haenszel measure estimates the extent to which the odds ratio deviates from 1. The Mantel–Haenszel estimator for the item is

$$
\alpha_{MH} = \frac{\sum_{k=1}^{s} n_{1rk} n_{0fk} / n_k}{\sum_{k=1}^{s} n_{0rk} n_{1fk} / n_k}
$$

where n_{1rk} is the number of persons at score level k belonging to the reference group and having a correct response to the item, and so forth. In practice, a logarithmic rescaling of α is used as a measure of effect size.

The Mantel–Haenszel chi-square statistic with 1 df (dropping a continuity correction) is

$$
\chi^2_{MH} = \frac{\left(\sum_{k=1}^{s} n_{1fk} - \sum_{k=1}^{s} En_{1fk} \right)^2}{\sum_{k=1}^{s} \mathrm{Var}(n_{1fk})}
$$

where

$$\mathrm{E}n_{1fk} = \frac{n_{1k}n_{fk}}{n_k}$$

and

$$\mathrm{Var}(n_{1fk}) = \frac{n_{1k}n_{0k}n_{rk}n_{fk}}{n_k^2(n_k-1)}$$

Both approaches are based on an analysis of the two-by-two tables of group membership and correct versus incorrect responses for the various total score levels. The Mantel–Haenszel statistic is easily extended to the case of more categories by replacing numbers correct by sums over persons of scores (see, e.g., Wang and Su, 2004).

Both approaches can be related to the IRT approach, with observed score as an indicator of latent ability. The Mantel–Haenszel method works well in case all items are Rasch items and the investigated item is the only biased item (Zwick, 1990).When the items are not Rasch items and the ability distributions of the focal and reference groups differ, one better takes another procedure for the detection of DIF. The procedure SIBTEST (Shealy and Stout, 1993) corrects for the fact that the raw score indicates different expected true scores in both groups.

When several items are biased, the detection of DIF becomes more difficult. The biased items have too much influence on the latent ability estimate. Therefore, Kok, Mellenbergh, and Van der Flier (1985) proposed an iterative method for the detection of DIF. Millsap and Everson (1993), Camilli and Shepard (1994), and Scheuneman and Bleistein (1999) give overviews of methods for the detection of item bias. Cole and Moss (1989) make some critical remarks with respect to DIF methodology and the interpretation of the outcomes of DIF studies.

Finally, an interesting approach related to DIF research is the mixed model approach proposed by Rost (1990, 1991). In this approach, a person does not belong to a group defined on the basis of an external criterion (e.g., female, male), but to a latent class. In each latent class, a unidimensional model is assumed to hold, but the item parameters differ between latent classes. The analysis might reveal the existence of classes of respondents who solve the test problems according to different strategies, resulting in different item difficulties. When there are two latent classes, a big one and a small one, the small latent class generates deviant response patterns, the subject of the next section.

10.5 Deviant answer patterns

Not only items may show deviations from the model specifications, but also persons can have responses that deviate from the pattern expected in the IRT model. An example is a low-scoring person who copies part of the answers of a high-scoring neighboring examinee (Wollack and Cohen, 1998). Wollack and Cohen, using the nominal response model, defined the following index for the similarity between the responses of an alleged copier and a source:

$$\omega = \frac{h_{CS} - \sum_{i=1}^{n} P(u_{iC} = u_{iS} \mid \theta_C)}{\sqrt{\sum_{i=1}^{n} P(u_{iC} = u_{iS} \mid \theta_C)[1 - P(u_{iC} = u_{iS} \mid \theta_C)]}} \qquad (10.4)$$

where n = number of items, C = copier, S = source, u_{iC} = response of C to item i, u_{iS} = response of S to item i, and h_{CS} = number of identical responses.

A high value of ω indicates that copying probably occurred. Several other copying detecting approaches were suggested. A recent suggestion by Sotaridona, Van der Linden, and Meijer (2006) is not based on IRT. More information on answer copying and cheating in general can be found in Cizek (1999).

Various general methods have been proposed for the detection of deviant answer patterns or the analysis of person fit, as it is also known in the literature.

Rudner (1983) suggested the $F2$ statistic as a generalization of a fit statistic used in connection with the Rasch model. The $F2$ statistic is

$$F2 = \frac{\sum_{i=1}^{n} [u_i - P_i(\hat{\theta})]^2}{\sum_{i=1}^{n} P_i(\hat{\theta})[1 - P_i(\hat{\theta})]} \qquad (10.5)$$

Drasgow, Levine, and Williams (1985; see also Drasgow, Levine, and McLaughlin, 1987) introduced z_3, the standardized ℓ_0 index for dichotomous items and a similar index for polytomous items, Index ℓ_0 is the logarithm of the likelihood evaluated at the maximum likelihood estimate of θ. An atypical response pattern is indicated by a low log likelihood of the response pattern $u_1...u_n$ given the estimated ability parameter (ℓ_0). But, what is a relatively low log likelihood? For an appropriate interpretation of the log likelihood, the expected value and the standard deviation of the log likelihood given the estimated person parameter are needed. The standardized index is

$$z_3 = \frac{\ell_0 - M(\hat{\theta})}{s(\hat{\theta})} \tag{10.6}$$

with

$$\ell_0 = \sum_{i=1}^{n} [u_i \ln P_i(\hat{\theta}) + (1 - u_i) \ln[(1 - P_i(\hat{\theta}))]] \tag{10.7}$$

The mean and variance of the log likelihood given the estimated value of θ are

$$M(\hat{\theta}) = \sum_{i=1}^{n} [P_i(\hat{\theta}) \ln P_i(\hat{\theta}) + [1 - P_i(\hat{\theta})] \ln[(1 - P_i(\hat{\theta}))]] \tag{10.8}$$

and

$$s^2(\hat{\theta}) = \sum_{i=1}^{n} P_i(\hat{\theta})[1 - P_i(\hat{\theta})]\{\ln P_i(\hat{\theta}) - \ln[1 - P_i(\hat{\theta})]\}^2 \tag{10.9}$$

Molenaar and Hoijtink (1990) noticed that ℓ_0 does not have a normal distribution. They proposed to use a chi-square distribution for the evaluation of deviant answer patterns. Another approach to the distributional problem is to generate the distribution. This approach was suggested by Glas and Meijer (2003) using the computationally intensive MCMC technique. Glas and Meijer obtained results for L_0 and several other indexes.

A review of many person fit indexes was given by Meijer and Sijtsma (2001).

10.6 Computerized adaptive testing (CAT)

Tests can be computer administered. A wide variety of item formats is available in computer-based tests, both items with a forced item response as well as items with open-response formats. Computerized testing makes it possible to allow different examinees to take a test on different occasions. For each examinee a different test can be composed, in order to avoid the risk that items become known among other things. Tests frequently are composed using a stratified random selection procedure. In that case, results can be analyzed with generalizability theory, and, when items are scored dichotomously, with approaches discussed in Chapter 6.

With computerized testing more is possible. It is possible, for example, to use a sequential testing design. One example of such an approach is the closed sequential procedure mentioned in Chapter 6.

With item response theory, computerized testing can be made even more flexible. First, consider a traditional test. Such a test is meant for measurements in a population of persons—the target population. No test can be equally accurate for all persons from the target population. However, with computerized adaptive testing, we have the possibility to administer each person a test in such a way that the test score is as accurate as possible.

If we knew the ability of a person, we could administer a test tailored to the ability level of this person. However, we do not know the ability level of a person; if we knew there would be no need for testing. Using item response theory, a testing strategy can be used such that step by step a person's ability is estimated. At each consecutive step, the estimate is more precise. The choice of the item or subset of items at each step is tailored to the estimated ability at the previous step. This calls for items for which item parameters have already been estimated. All the items are stored in an item bank, and for this large set of items IRT item parameter estimates are known.

More technically, for the administration of the first item, we can start with the not unreasonable assumption that a person to be tested is randomly chosen from the target population. The population distribution can be regarded as the prior distribution for the ability of this person. After each response, we can compute the posterior distribution of θ from the prior distribution $g(\theta)$ and the likelihood of all responses $L(\mathbf{x} \mid \theta)$ (cf. Equation 9.29). This distribution can be used as the new prior distribution for the ability of the person. We choose a new item optimal with respect to this prior. We might, for example, after a response to an item, compute the posterior mean, the EAP estimate, and select a new item that has the highest item information at the level of the EAP estimate. After a correct response, the estimated ability is higher than after an incorrect response. Therefore, a more difficult item is administered after a correct response than after an incorrect response. We might stop when the error variance is smaller than some criterion. When the EAP estimate is used, the relevant error variance is the posterior variance of θ (Bock and Mislevy, 1982). This CAT procedure is illustrated in Figure 10.3. For practical reasons, another stopping rule is frequently used: the test length of the CAT procedure is fixed. Test length is variable, however, in applications where a decision about mastery must be made (i.e., when it must be decided whether an examinee has an ability level equal to or higher than a minimum proficiency level θ_c) (Chang, 2005).

Sometimes it is profitable to redefine the unit of presentation in CAT and to group items into testlets. One argument for grouping could be that several items are based on the same subject or the same text, but there might be other reasons for grouping items as well (Wainer and Kiely, 1987). With a redefinition of the unit of presentation, a different choice of item response model might be in order (Li, Bolt, and Fu, 2006; Wainer and Wang, 2000).

Another approach to CAT is exemplified by the ALEKS software used for learning in highly structured knowledge domains like basic math (www.aleks.com). An early psychometric contribution to this approach is by Falmagne (1989). Another contribution to flexible testing has been made by Williamson, Almond, Mislevy, and Levy, (2006).

Computerized adaptive testing (CAT) can be very efficient in comparison to traditional testing (Sands, Waters, and McBride, 1997; Van der Linden and Glas, 2000). With a relatively short test length, we already obtain a highly accurate ability estimate. This removes the objection to the use of a prior distribution. With an accurate test, the weight of the prior in the final ability estimate is very small. CAT can also be used in connection with multidimensional traits. Li and Schafer (2005) discuss multidimensional CAT in which the items have a simple structure (i.e., load on only one of the latent trait dimensions).

In practice, concessions have to be made in order to make CAT feasible. If we would use only items with maximum information given the estimated ability, then we would probably use a limited number of items from a large item pool frequently and other items would never be used. Several methods have been proposed to deal with this problem (Revuelta and Ponsoda, 1998). Van der Linden and Veldkamp (2004) used the concept of shadow tests for constraining item exposure.

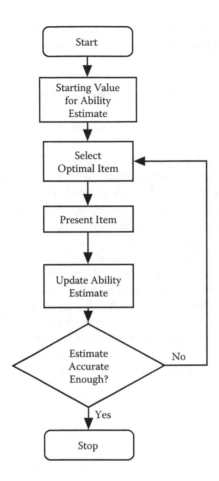

Figure 10.3 Flowchart of computerized adaptive testing (CAT) with a stopping rule based on estimation accuracy.

When CAT is to be introduced, a few aspects of testing with CAT must be attended to. Answering the CAT test is different from answering a traditional test. The items must be answered consecutively; skipping items is not allowed. Therefore, it is sound practice to study the validity of the test procedure. We should be alert to the possibility that the validity of the test changes with a change in procedure.

The interest in CAT is growing tremendously, especially because of its prospects in educational assessment. The future of testing will be determined, among others, by CAT (see, e.g., *Standards*, APA et al., 1999).

In 1995 the American Council on Education published the *Guidelines for Computer-Adaptive Test Development and Use in Education.*

10.7 IRT and the measurement of change

The measurement of change is beset with problems:

- The first problem has to do with measurement error that gives rise to the phenomenon of the regression to the mean.
- The second problem has to do with the question of whether change can be interpreted as change along one dimension—can scores before and after a change be interpreted in terms of the same underlying construct?
- The third problem has to do with the limitations of the measurement instruments used. When the standing of a person on a latent scale increases, a higher score can be expected on the measurement instrument. A person with an intermediate score can have a large observed score gain. A person with a relatively high initial score cannot have a large observed gain: There is a maximum score on the test. In the measurement of change, a ceiling effect is to be expected.

It is clear that IRT cannot solve all problems associated with the measurement of change. Using estimated scores on a latent scale defined by an IRT model can at least eliminate the problem of ceiling effects. An early proposal with respect to the use of IRT for the measurement of change was made by Fischer (1976). Fischer did more than propose to use the latent ability scale for measuring change, he also proposed to model the amount of change for each individual as a weighted sum of effects. Other models for the measurement of change over several occasions have been proposed by Embretson (1991) and Rijmen, De Boeck, and Van der Maas (2005).

Here, we present only the simplest linear logistic test model discussed by Fischer. Let us assume that from time t_1, several treatments are given. At time t_1 the dichotomous Rasch model holds with abilities θ_p and item parameters b_i. At time t_2 the probability of a correct response is described by

$$P_i(\theta_p \mid t_2) = \frac{\exp(\theta_p + \delta_p - b_i)}{1 + \exp(\theta_p + \delta_p - b_i)} \qquad (10.10)$$

with

$$\delta_p = \sum_{j=1}^{m} q_{pj}\eta_j + \tau \qquad (10.11)$$

where q_{pj} is the "dose" of treatment j as applied to person p, η_j is the treatment effect for persons who took the whole treatment, and τ measures a general trend or development independent of the treatments. In the simplest case, the dose is 0 or 1, with 0 designating no treatment. Persons who took the same combination of treatments show the same change δ. An alternative interpretation is that the item difficulties have decreased by the same amount δ in this group of persons. So, in the subgroup of persons with the same combination of treatments, the n items at the first test administration together with the same n items at the second test administration must conform to the Rasch model.

10.8 Concluding remarks

The state of the art at the end of the 20th century is given in the *Handbook of Modern Item Response Theory* (1997), edited by Van der Linden and Hambleton. Not only are many IRT models presented by experts in the field, but also examples are provided (although limited in number). Most of these models have been included in the previous chapter of the present monograph. Interestingly, the models can be classified according to roughly the following criteria:

- Response format (dichotomous, polytomous items; ordered versus unordered categorical data, open-end items)
- Response time or number of successful attempts as responses on test items
- Unidimensional or multidimensional items in a test
- Type of response function (monotonous versus nonmonotonous; type of the response function [e.g., normal ogive, logistic, hyperbolic cosine, Cauchy])
- Single versus multiple-group analysis.

Combining these five criteria maps a whole gamut of IRT models, giving work for a whole army of research workers for decades to come. And if the application of IRT models is taken seriously, then not only is cooperation between basic and applied researchers a prerequisite, but also a selection of the applied fields necessary (e.g., performance assessment, test fairness, setting standards, certification, and the like in educational testing; the measurement of human abilities, measurement in personality, clinical and health psychology, developmental psychology, attitude measurement, personnel psychology). In the context of applied measurement, more attention should be paid to the interpretation of model parameters, in addition to more technical matters as model identification and parameter estimation.

But where are we now? What are the achievements and blessings of IRT? Is it a fair assessment by, for example, Goldstein and Wood (1989) or Blinkhorn (1997) that, to rephrase Horace, a mountain of IRT models gave birth to a silly little mouse of insight?

IRT has led to some fruitful results in the field of equating or research on the comparability of measures, on fairness in testing and test use (DIF research), and last but not least computerized adaptive testing (CAT). Specifically, in the field of CAT, IRT is indispensable. One of the major developments of educational and psychological testing is CAT, and CAT is nigh to impossible without taking refuge in IRT. Daniel (1999) argues that IRT is indispensable for improving the adaptive administration of intelligence tests.

It is not all roses, however, with IRT. One class of problems with IRT is methodological and technical in nature. For example, what does it really mean when assumptions are violated, and then, how to proceed? Surely, when the unidimensionality assumption is violated, we can take refuge in multidimensional IRT. But what is the nature of those multiple dimensions? How do we interpret them? How stable are the results on the item information function under deviations of ICCs from the normal ogive and logistic models (see Bickel, Buyske, Chang, and Ying, 2001)? What does IRT contribute to test validation? Problems of a technical nature have to do with estimation of model parameters, and with model testing. The three-parameter logistic model, for example, requires sample sizes of at least 1000 for a moderate number of items (say 40) in a test to achieve stable parameter estimates. And with more elaborated models, the estimation problems become more complicated.

A second class of problems has to do with the implementation of IRT, for example, for personality assessment. In personality assessment, a wide variety of constructs exist. Can all these constructs be modeled appropriately by an IRT measurement framework? Reise (1999, pp. 237–238) argues that most available IRT models are too restrictive for personality assessment.

Of course, it is no excuse of applied research workers and practitioners in testing to shun the use of IRT procedures because these are tedious to apply and difficult to understand. On the other hand, research workers would be ill advised if all should climb on the IRT bandwagon and leave classical and neoclassical (i.e., generalizability) test theory behind. These major test theories and their corresponding procedures for test development, validation, and evaluation must continue to exist side by side, but cross-fertilization should be enhanced.

Exercises

10.1 Two test items were administered to a reference group R and a focal group F. The proportions correct are

$p_{1(R)} = 0.70$

$p_{1(F)} = 0.65$

$p_{2(R)} = 0.70$

$p_{2(F)} = 0.60$

Is the second item biased against the focal group?

10.2 We have five Rasch items with $b_1 = -0.5$, $b_2 = -0.3$, $b_3 = 0.0$, $b_4 = 0.25$, and $b_5 = 0.5$. We want to construct a two-item test that discriminates relatively well at $\theta_1 = -0.5$ and at $\theta_2 = 0.5$. Which combination of two items from the item bank with five items is best, given the criterion that the largest of the error variances at θ_1 and θ_2 should be as small as possible?

10.3 In an item bank we have items conforming to the 2PL model. The items have the item parameters: $b_1 = -0.5$, $a_1 = 1.0$, $b_2 = -0.25$, $a_2 = 2.0$, $b_3 = 0.0$, $a_3 = 0.7$, $b_4 = 0.25$, $a_4 = 1.0$, $b_5 = 0.5$, and $a_5 = 1.5$. We test a person and the present point estimate of ability θ is 0.20. Which item should be presented next to this person?

CHAPTER 11

Test Equating

11.1 Introduction

In many test situations, multiple forms of a test are made available to assess ability, achievement, performance, or whatever. When persons are administered several test forms meant to measure the same ability, we want to be able to compare these persons' test scores. With parallel tests this can be done straightforwardly. Parallel tests measure the same content and share statistical specifications (equal means, standard deviations, and reliabilities). That is to say, scores on parallel tests are completely exchangeable. No comparison problem occurs with parallel forms of a test. More often than not, multiple forms of a test that measure the same attribute are not parallel, and a comparison of scores is not straightforward, because test forms may differ in several respects (unequal means, unequal variances, unequal reliability, and the like). So, before comparing persons' or examinees' scores on multiple forms of the same test, it is necessary to establish, as nearly as possible, an effective equivalence between raw scores on the multiple forms of a test. This is the problem of equating, a problem that is different from the problem of developing test norms as elucidated in Exhibit 11.1.

Exhibit 11.1   Equating and norming

For many test applications, it is necessary to know the distribution of test scores in a particular population. This means that norms for the test have to be developed. Norming means that an adequate sample from the population is approached, most frequently using stratified sampling from the relevant population. Two different tests supposed to measure the same characteristic can be normed for the same population. Can these two tests also be equated on the basis of the norming data?

When different researchers construct tests measuring, say, intelligence, their conception of what intelligence entails may differ. The constructs that are measured are likely to be dissimilar. Researchers also may use different sampling plans, and for this reason, norms cannot be equated. Finally, norms obtained at different times can differ because of changes in the characteristics of the relevant population. Flynn (1987; see also Dickens and Flynn, 2001) showed that there have been large gains in scores on intelligence tests over time. Such changes imply that old norms become obsolete after some time and new norms have to be developed.

Before going into the equating problem, we must keep in mind that in general the process of associating numbers with the performance of persons or examinees on tests is called scaling. This process leads to scale scores (raw scores, normalized scores, stanines, and the like). The process of scaling must be distinguished from the process of equating. Equating procedures are used to insure that scores from the administration of multiple forms of a test can be used interchangeably. Test theorists and practicians differ in opinion, however, on what conditions should be met for equated scores (i.e., the scores obtained after applying equating methods). Not only can interchangeability refer to alternative and weaker forms of strict parallelism of measurement instruments, as discussed in Chapters 3 and 4, but also to test content and to the target population for which the test is intended. To be more precise, the following four conditions or properties of equated test scores are pertinent:

1. Same ability (i.e., alternative test forms must measure the same characteristic—ability, achievement, or performance)
2. Equity (i.e., for every group of persons or examinees given the same ability, the conditional frequency distribution of scores on one of the test forms, say test Y, is the same after transformation as the conditional frequency distribution of untransformed scores on the other test, test X)
3. Population invariance (i.e., the transformation is the same irrespective of the sample or group of persons from which it is derived)
4. Symmetry (i.e., the transformation is reversible; transforming the scores of form X to form Y is the same as transforming the scores of form Y to form X)

The explicit definition of condition 2 is given by Lord (1980, Chapter 13). If complete equity after equating or transformation of scores on test forms X and Y is observed, then both forms of the test are strictly parallel in the sense of classical test theory. Complete equity according to the definition by Lord is hardly feasible in practice, be it for the simple reason that very often reliabilities differ. Low-ability examinees have an advantage with a relatively low reliability, whereas high-ability examinees have an advantage with an accurate measurement of their ability, in other words, with a relatively reliable test. So, it should be clear how important it is to make tests as comparable as possible with respect to reliability.

After equating, at least the expected score or true score on one test should be equal to the true score on the other test. In terms of true scores of two tests X and Y, we should obtain

$$T'_Y = T_X$$

in other words, after equating test Y with test X the true score on test Y is equal to the true score on test X. As already mentioned, meeting all four conditions or desirable equating properties is nigh to impossible. In actual equating practice, they can only—hopefully as close as possible—be approximated. And as investigators in testing programs differ in opinion on what closeness of approximation entails, one or more of the conditions or properties are aimed at. Specifically, the conditions or desirable properties of equated scores are discussed in more detail by Lord (1980), Kolen (1999), and Petersen, Kolen, and Hoover (1989).

Test equating is an empirical enterprise. It boils down to establishing a relationship between raw scores or scale scores in general on two or more test forms: data on multiple test forms have to be collected, and then appropriate equating methods have to be applied for transforming the scores. In Section 11.2 three basic equating designs for collecting data are outlined. In Section 11.3, equipercentile equating is introduced, in Section 11.4 linear equating. Linear equating methods that make use of an anchor test are presented in Section 11.5. Section 11.6 introduces the kernel method of observed score equating. In Section 11.7 IRT-based equating methods are presented on an elementary level, without losing the gist and flavor of them, however. In the final Section 11.8 some concluding comments are made.

We mention some of the important publications here. Angoff (1971, 1984 published as a separate monograph) gave one of the first extensive treatments, a chapter in *Educational Measurement* (second edition). In the third edition of *Educational Measurement*, Petersen, Kolen, and Hoover (1989) coined their contribution "Scaling, norming, and equating." There is a special issue of *Applied Psychological Measurement*

(Brennan, 1987), and a special issue of *Applied Measurement in Education* (Dorans, 1990) to give overviews of research and development in this field. More specifically, Skaggs and Lissitz (1986) review IRT equating, and Cook and Petersen (1987) discuss problems related to the use of conventional and IRT equating in less-than-optimal circumstances. One of the books on equating is the monograph by Kolen and Brennan (1995). There is another book, by Von Davier, Holland and Thayer (2004), in which a unified approach to observed test equating is proposed. Last, but not least, the new *Standards for Educational and Psychological Testing* (APA et al., 1999) should be mentioned. The importance and relevance of the equating of tests is exemplified by including a special chapter in the *Standards* on scales, norms, and score comparability or equating.

11.2   Some basic data collection designs for equating studies

There are several methods that can be used to equate scores on multiple test forms. These equating methods are tuned to the particular data collection design. Here we will discuss only three basic data collection designs. A more extended list, including section pre-equating (Holland and Wightman, 1982) can be found in Petersen et al. (1989).

11.2.1   Design 1: Single-group design (Figure 11.1a)

In this design, forms X and Y are both given to one group of persons or examinees. A disadvantage of this design is that much time is needed for the administration of the tests. Fatigue might play a role when persons answer the items of the second test. Therefore, the best thing to do is to administer the tests in a different order to part of the persons. Technically speaking, this design is a counterbalanced random-groups design. The single group is split into two random-half samples and both half-samples are administered test forms X and Y in counterbalanced order (e.g., the first half-sample takes form X first, while the second half-sample takes form Y first). This is realized by administering the tests in rotation to a group of persons who are present at the test session. When the design is the counterbalanced random-groups design, there are several alternative estimation approaches.

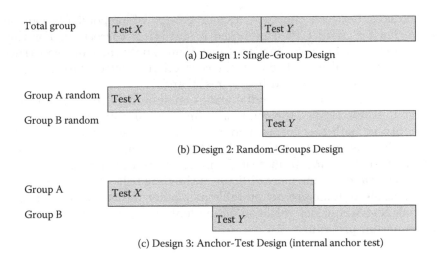

Figure 11.1 Designs for obtaining data in equating studies.

11.2.2 Design 2: Random-groups design (Figure 11.1b)

In this design, test forms X and Y are administered to different random samples from the population.

With large-scale examinations, one of the tests, say test X, is the old test and the other test, test Y, is a new test to be equated to the old test. An equating study using Design 1 or Design 2 is not possible in this situation because the contents of the new test would become available prematurely. Design 3 does not have this disadvantage.

11.2.3 Design 3: Anchor-test design (Figure 11.1c)

In this design, all persons are given a test V that is functionally equal to tests X and Y. So, test X and test V are administered to one sample of persons, and test Y and test V to another sample of persons. The two samples may differ from each other in a nonrandom way. The common test V is called the anchor test. Test V may be a common subtest of test forms X and Y; in that case, we talk of an internal anchor test. Test V might also be a third test in which case we have an external anchor test. Tests X and Y can be related to each other by means of the common or anchor test V.

Design 3 can also be used with two random samples of persons. An advantage of Design 3 in comparison to Design 2 is that eventual differences between the two random groups of respondents can be corrected for. Lord (1955) proposed a statistical correction. For this correction, it is not necessary that test V measures the same construct as tests X and Y, but, of course, the correction is better with a higher correlation between V and test forms X and Y.

The items from an internal anchor test should, of course, be insensitive to context. Also, the common items should have the same functionality in both test forms X and Y (i.e., the items need to behave similarly). Assume that test X has been administered a long time before test Y. Items from test X might have become obsolete and therefore have become more difficult at the time test Y is administered. Such obsolete items are not suitable as anchor items: inclusion in the anchor test would not result in equivalent scales after application of an equating procedure.

11.3   The equipercentile method

The property of population invariance of equating can only be approximated in practice, especially when raw scores on two test forms are used. So, it is important to define the population for which the relationship between two tests X and Y has to be obtained. In equipercentile equating, raw scores on test forms X and Y are considered to be equated if they correspond to the same percentile rank in the population.

Suppose that we administered two forms X and Y of a test to a large group of persons from the relevant population (Design 1). The tests have the same reliability and there are no context effects. Then two scores x and y are equivalent—apart from measurement error—if the two scores have an identical percentile rank, in other words, if equal percentages of persons have these scores or lower scores on the tests. With Design 2, in principle, the same definition of score equivalence can be used. The difference with Design 1 is that sample fluctuations introduce more error in the estimated relationship between the tests. The equating process for equating with the equipercentile method is demonstrated with Table 11.1 and Figure 11.2 and Figure 11.3.

In Table 11.1 the percentile scores of two 40-item test forms X and Y are given. The percentile score associated with a particular raw score y on test form Y equals the percentage of persons with score $y - 1$ or a lower score plus half the percentage of persons with score y. In the table, we find that raw score 20 on form Y corresponds to a percentile score equal to 30.3. In test form X, raw score 20 corresponds to a percentile score equal to 45.3. The score on test form X that is

Table 11.1 Percentile scores of two test forms X and Y, and scores X equated to scores Y.

Raw Score	Percentile Score Y	Percentile Score X	Score X Equated to Y	Raw Score	Percentile Score Y	Percentile Score X	Score X Equated to Y
0	0.0	0.0	0.0	21	35.3	51.6	18.3
1	0.0	0.0	1.0	22	40.6	57.5	19.3
2	0.0	0.0	2.0	23	45.3	63.9	20.0
3	0.0	0.0	3.0	24	50.7	69.8	20.9
4	0.0	0.0	4.0	25	56.9	74.2	21.9
5	0.0	0.1	4.0	26	62.4	78.0	22.8
6	0.0	0.2	4.0	27	67.2	81.4	23.6
7	0.2	0.5	6.0	28	72.2	85.0	24.5
8	0.6	1.2	7.1	29	77.1	88.3	25.8
9	1.0	2.0	7.7	30	81.7	90.8	27.1
10	1.5	3.3	8.3	31	85.7	93.0	28.2
11	2.2	5.0	9.2	32	89.5	95.6	29.5
12	3.3	6.8	10.0	33	92.8	97.3	30.9
13	4.7	9.2	10.8	34	95.2	98.3	31.9
14	6.5	12.7	11.8	35	97.5	99.2	33.2
15	9.1	16.9	13.0	36	99.1	99.5	34.9
16	12.2	21.6	13.9	37	99.6	99.9	36.3
17	15.9	27.5	14.8	38	99.7	100.0	36.5
18	20.7	33.4	15.8	39	99.9	100.0	37.0
19	25.8	39.0	16.7	40	100.0	100.0	40.0
20	30.3	45.3	17.5				

equivalent to score 20 on test form Y must have the same percentile score: 30.3. There is no raw score on X with this percentile score. The equated score on X must have a value between 17 and 18. Linear interpolation gives the value 17.5 as the equated score on test X. In this way, we can find equated scores x for all scores y. Those values are given in the table. Not in the table are the scores y equated to the raw scores 0 up to and including 40 on test form X.

The procedure of equipercentile equating of the raw scores on test forms X and Y is depicted in Figure 11.2. In Figure 11.3 the obtained relation between scores on test form X and scores on form Y is displayed.

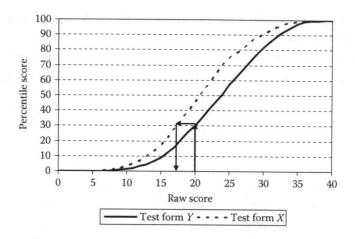

Figure 11.2 The percentile scores of test forms X and Y, and the construction of equated scores.

Equipercentile equating is extremely sensitive to sampling fluctuations. This is especially the case at the low and high ends of the score scales where the computation of percentile scores rests on small numbers of observations. Several approaches have been suggested to diminish the influence of sampling fluctuations. All methods are based on smoothing. In presmoothing methods, the score distributions of x and y are smoothed before equating the tests. In postsmoothing, the obtained

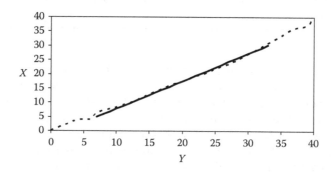

Figure 11.3 The equated scores on forms X and Y, obtained with the equipercentile method; also a linear approximation of the relation between the scores on X and Y is shown.

relationship between X and Y is smoothed (Kolen and Brennan, 1995, Chapter 3).

When test forms X and Y are approximately linearly related, the obvious smoothing technique is to equate tests by means of a linear equating method. As Figure 11.3 shows, linear equating might be adequate for a large range of scores.

The equipercentile method for equating can be applied to data obtained with all the designs discussed earlier in Section 11.2. How equating proceeds in practice is given by Kolen and Brennan (1995), among others. Also, standard errors of equipercentile equating are presented there.

11.4 Linear equating

If tests have about the same score distribution apart of their means and standard deviations, a linear equating method is sufficient. In linear equating, a transformation is chosen such that scores on two test forms are considered equated if they correspond to the same number of standard deviations above or below the mean in the same group of persons. Because of the typical character of linear equating, we must take some measures for equating at high and low scores because linear equating inevitably leads to impossible transformed scores, negative scores, and scores higher than the maximum score. For linear equating in Design 1 or Design 2, the following equation is used:

$$\frac{x - \bar{x}}{s_X} = \frac{y - \bar{y}}{s_Y} \tag{11.1}$$

where x and y refer to the scores on the test forms to be equated. Using Equation 11.1, we can write the score on Y after transformation to the scale of X, y', as

$$y' = Ay + B \tag{11.2}$$

where

$$A = \frac{s_X}{s_Y} \tag{11.3}$$

and

$$B = \bar{x} - A\bar{y} \tag{11.4}$$

In Design 2, two random samples of persons are needed. Test form X is administered to one of these samples, group A. The mean and standard deviation of scores on X are obtained from this group. Test Y is administered to the second sample, group B. Group B gives a mean and standard deviation for scores on Y. If groups A and B cannot be regarded as random samples from the same population, equating is not possible with Design 2. An anchor test is needed to make a correction for group differences possible, in which case the data collection design for linear equating of scores is Design 3.

11.5 Linear equating with an anchor test

In Design 3 we have an internal or external anchor test V that has the same function as test forms X and Y. By means of the common anchor test, tests X and Y can be equated. First, we must define the population for which the equating relation is to hold. This population, the so-called *synthetic population*, might be defined by combined group A + B, but other population definitions are possible (Kolen and Brennan, 1995, pp. 106–111). Using data for test X and V, and test Y and V, we can estimate means and standard deviations of X and Y in the synthetic population. Next, Equation 11.2 through Equation 11.4 are used to define the equivalence relationship between tests X and Y. Several methods for estimating means and standard deviations of tests X and Y in the synthetic population are available. Tucker proposed a method for linear equating that can be used if groups A and B do not differ much in ability. The resulting equating equation is formally identical to a result obtained by Lord under the assumption of random groups A and B. Levine developed two methods for samples that may differ widely in ability. The first method can be used with equally reliable tests. The second method is suitable when the test forms differ in reliability. In the second case, the true scores can be scaled to the same scale, but obviously raw scores cannot be equated.

The first method is described in Exhibit 11.2. The second method is easier to derive than the first Levine method or the Tucker method (for linear observed score equating approaches, see Von Davier and

Kong, 2005). Due to the fact that the second Levine method is defined for true scores, the computation of means and standard deviations of test scores for the synthetic population can be avoided.

Exhibit 11.2   Levine's first method: Equally reliable tests

With equally reliable tests, the mean and standard deviation of test forms X and Y are estimated for a synthetic group. Here, we estimate the mean and variance of X and Y for the total group T = A + B. The procedure is illustrated for test form X.

Two assumptions are made with respect to test form X and the common test V. The first assumption is that the true scores of X and V are linearly related. This assumption leads to two equations:

(a)
$$\overline{T}_{X(\mathrm{T})} - \frac{s_{T_X(\mathrm{T})}}{s_{T_V(\mathrm{T})}}\overline{T}_{V(\mathrm{T})} = \overline{T}_{X(\mathrm{A})} - \frac{s_{T_X(\mathrm{A})}}{s_{T_V(\mathrm{A})}}\overline{T}_{V(\mathrm{A})}$$

and

(b)
$$\frac{s_{T_X(\mathrm{T})}}{s_{T_V(\mathrm{T})}} = \frac{s_{T_X(\mathrm{A})}}{s_{T_V(\mathrm{A})}}$$

that is, the intercept and the slope of the relation of the true scores of X and V is the same in groups A and T.

The second assumption is that the variance of measurement errors for test form X is the same in group A and group T,

(c)
$$s_{X(\mathrm{T})}^2(1 - r_{XX'(\mathrm{T})}) = s_{X(\mathrm{A})}^2(1 - r_{XX'(\mathrm{A})})$$

Using (a) and (b) and substituting the observed mean for the true score mean, we obtain

$$\overline{x}_{\mathrm{T}} = \overline{x}_{\mathrm{A}} + \frac{s_{T_X(\mathrm{A})}}{s_{T_V(\mathrm{A})}}(\overline{v}_{\mathrm{T}} - \overline{v}_{\mathrm{A}})$$

Using (b) and (c), we obtain

$$s^2_{X(T)} = s^2_{X(A)} + \frac{s^2_{T_X(T)}}{s^2_{T_V(T)}}\left(s^2_{V(T)} - s^2_{V(A)}\right)$$

The observed-score variance of test form X in the total group can be obtained when the ratio between the true-score variance of X and the true-score variance of V is known. An estimate of this ratio is presented in the main text.

Next, the mean and variance of test form Y in the total group are estimated. Finally, Equations 11.2 through 11.4 are used to equate test forms X and Y.

In the second Levine method, it is assumed that the true scores on X and V are linearly related, and similarly that the true scores on Y and V are linearly related. A true score on test X, T_X is equivalent to a true score on test V, T_V if the two scores have the same z-score within the same group, say group A, of persons:

$$\frac{T_X - \mu_{T_X(A)}}{\sigma_{T_X(A)}} = \frac{T_v - \mu_{T_V(A)}}{\sigma_{T_V(A)}} \tag{11.5}$$

This equation can be rewritten as follows:

$$T_X = \gamma_{XV}(T_V - \bar{v}_{(A)}) + \bar{x}_{(A)} \tag{11.6}$$

where observed-score means are substituted for true-score means. In the equation, γ_{XV} denotes the ratio between the true-score standard deviation on X and the true-score standard deviation on V. This ratio is assumed to be group independent. A similar equation can be obtained for the relation between true scores on Y and true scores on V. Coefficients for the equation relating Y and V can be obtained from group B:

$$T_V = (T_Y - \bar{y}_{(B)})/\gamma_{YV} + \bar{v}_{(B)} \tag{11.7}$$

Substitution of T_v from Equation 11.7 in Equation 11.6 produces the following relationship between the true-score scales of tests X and Y:

$$T_X = \frac{\gamma_{XV}}{\gamma_{YV}}[T_Y - \bar{y}_{(B)}] + \gamma_{XV}[\bar{v}_{(B)} - \bar{v}_{(A)}] + \bar{x}_{(A)} \qquad (11.8)$$

Next, raw scores on X and Y are equated as if they were true scores. The correction for the difference in ability level between groups A and B is one of the differences with Equation 11.2 through Equation 11.4. The second difference has to do with γ. Angoff (1971) called the ratio of true-score standard deviations γ effective test length. He assumed that test X can be regarded as a combination of γ_{XV} tests parallel to anchor test V. Similarly, test Y might be regarded as a test composed of γ_{YV} tests parallel to test V. This is a stronger requirement than the requirement that the three tests X, Y, and V have linearly related true scores. With Angoff's assumption, the coefficients γ can easily be determined. In case test V is included in test X, γ_{XV} is computed as follows:

$$\gamma_{XV} = \frac{s_x}{s_v r_{xv}} \qquad (11.9)$$

In this case, factor γ equals the inverse of the regression coefficient for the regression of V on X (Angoff, 1953). The coefficient is estimated from responses in group A. The factor γ_{YV} is estimated from the responses in group B. If V is an external test, another equation than Equation 11.9 is needed. Standard errors for the Levine procedure are given by Hanson, Zeng, and Kolen (1993).

When the common test V does not have the same function as test forms X and Y, equating is possible when the two groups A and B are random samples, using the method proposed by Lord (1955). If this is not the case, equating is not possible, but it is still possible to obtain comparable scores for test forms X and Y. Scores on X and Y might be defined as *comparable* if they are predicted by the same score on V. The definition of comparable scores as scores that are predicted by the same score on a third test is not the only definition possible. There are other definitions of comparability. The issue of comparability of scores is discussed at some length by Angoff (1971, pp. 590–597).

11.6   A synthesis of observed score equating approaches: The kernel method

Previously, it was argued that linear equating is an approximation to equipercentile equating. The relationship between linear and nonlinear approaches to observed score equating has been described in more precise terms by Von Davier et al. (2004). These authors proposed a general approach to equipercentile-like equating with linear equating as a special case. They start with presmoothing data using log-linear modeling. Next, they estimate the score distributions in the synthetic population on the tests X and Y that are to be equated. Gaussian kernel equating is used to transform these discontinuous distributions into continuous distributions, which makes linear interpolation obsolete. The results depend on the value of a "bandwidth" parameter, with large values leading to linear equating. In the next step, the equating function is obtained. Finally, standard errors of equating are computed.

11.7   IRT models for equating

Instead of the equipercentile method or the linear method of equating, a method that is based on IRT can be used. Equating with IRT has a large advantage over equating with the classical approach. With an IRT model that fits, the nonlinearities inherent in equating do not present a problem. IRT models can be used in *horizontal equating* as well as in *vertical equating*. In horizontal equating, different tests are meant for persons of similar abilities; equating as discussed so far is horizontal equating. In vertical equating, tests are constructed for target groups of different ability levels. The difference in test difficulty is planned, but for score interpretation, scores should be brought to the same scale. It is still necessary that all items be relevant for all examinees. Equating is not achieved if younger examinees have not been exposed to material tested in the unique items of the higher-level test tailored to the ability of a group of older examinees (Petersen et al., 1989). It should also be clear that in vertical equating, tests are not equated in the sense that they may be used interchangeably after equating.

In principle, three equating approaches for two test forms X and Y sharing a common set of items are possible within the IRT context (Petersen, Cook, and Stocking, 1983):

A. Simultaneous scaling: The item parameters of both tests are estimated jointly in one analysis. For this approach, we need software that allows for incomplete data—each person has answered only a subset of all items.

B. The responses to tests X and Y are analyzed separately. In the analysis of the second test, the item parameters of the common items are fixed to their values obtained in the analysis of the first test. The scales of X and Y can be related to each other by means of the scale values of the common items.

C. The responses to tests X and Y are analyzed separately. The difference with approach B is that the parameter values of the common items are not fixed to their values obtained in the analysis of the first test. Again, the scales of tests X and Y can be related to each other by means of the scale values of the common items.

When approach A is chosen and MML is the estimation method, characteristics of the latent ability distributions involved should be allowed to differ. Alternative C seems easiest to implement. Let us consider this approach in the context of the three most popular IRT models: the Rasch model, 2PL model, and 3PL model.

11.7.1 The Rasch model

With the Rasch model, the third approach is very straightforward. We need the averages of the b parameters of the common items in test X and in test Y. Suppose that we have k common items with the following averages:

$$\bar{b}_{X(c)} = \frac{1}{k}\sum_{i=1}^{k} b_{iX(c)}, \quad \bar{b}_{Y(c)} = \frac{1}{k}\sum_{i=1}^{k} b_{iY(c)} \tag{11.10}$$

The b parameters of both tests would be on a common scale if the average parameter value for the common items would be equal for both tests. So, the b values and θ values of test Y can be brought on the same scale as those of test X with the following transformation:

$$b_i^* = b_i - \bar{b}_{Y(c)} + \bar{b}_{X(c)}, \quad \theta^* = \theta - \bar{b}_{Y(c)} + \bar{b}_{X(c)} \tag{11.11}$$

We do not have the parameter values of the difficulty parameters, but only estimated values. The estimated values are not equally accurate. So we might consider using weighted averages instead of the unweighted average in Equation 11.11. Such a method has been proposed by Linn, Levine, Hastings, and Wardrop (1981).

11.7.2 The 2PL model

In the 2PL model, equating is a bit more complicated because the parameters are defined on an interval scale. The common items can have different a values as well as different b values. Because of the interval character of the latent scale, b parameter values of the common items of test Y are linearly related to the values for test X:

$$b_{i(X)} = db_{i(Y)} + e \qquad (11.12)$$

and the values of the a parameters are related through

$$a_{i(X)} = a_{i(Y)}/d \qquad (11.13)$$

The coefficients d and e must be obtained in order to bring the parameters of the common items, and consequently the parameters of all items, to the same scale.

The simplest solution is to find the transformation by which the average b value of the common items and the standard deviation of the b values of the common items are equal in both tests. This is the *mean and sigma* method. With this method, the value of d is

$$d = \frac{s_{b_{X(c)}}}{s_{b_{Y(c)}}} \qquad (11.14)$$

that is, the ratio of the standard deviation of the common b values in test form X to the standard deviation of the common b values in test form Y, and the value of e is

$$e = \bar{b}_{X(c)} - d\bar{b}_{Y(c)} \qquad (11.15)$$

A robust alternative is the previously mentioned weighted method.

We have two sets of parameter estimates for the common items. One set is computed along with the other item parameters in test X. The other set is computed along with the other item parameters in test Y. We also can compute two test characteristic curves—the sums of the ICCs of the items in the tests—for the subset of common items. After test equating, these two test characteristic curves should be similar. In the characteristic curve methods (Haebara, 1980; Stocking and Lord, 1983), coefficients d and e are obtained for which these test characteristic curves are as similar as possible.

11.7.3 The 3PL model

The 3PL model is also defined on an interval scale, but the presence of a pseudo-chance-level parameter c complicates the equating of tests. When we analyze two tests X and Y separately, the estimated c of a common item can have one value in test Y and another in test X. This difference is related to differences in the other parameter estimates of the particular item. The errors in the pseudo-chance-level parameters can have a disturbing effect on the relationship between the b parameters and the a parameters of the common items. In other words, we may expect a disturbing influence on the linear relationship between the item difficulties in test X and those in test Y. Equating tests X and Y is not simply achieved by using a linear transformation for the b values of the common items. With the 3PL model, more steps are needed. In a preliminary analysis, we obtain estimates of the parameters c. For a common item, one value for c is chosen on the basis of the two different values obtained in the analyses of tests X and Y. The chosen value can be the average of the two estimates. In the final analysis, the c parameter of a common item is fixed to this value for both tests.

After scaling the two test forms on a common latent scale, the relation between the true scores of both test forms can be computed. For each value of θ, the corresponding true score of test form X and the corresponding true score of test form Y can be computed:

$$\tau_X = G(\theta) = \sum_{i \in X} P_i(\theta) \tag{11.16}$$

and

$$\tau_Y = H(\theta) = \sum_{i \in Y} P_i(\theta) \tag{11.17}$$

The two true scores corresponding to the same θ are equated with the following formula:

$$\tau_X = G(H^{-1}(\tau_Y)) \tag{11.18}$$

that is, we take the true score on test form Y, compute the corresponding value of θ, and, next, compute the true score on form X for this value of θ. True-score equating does not work in the 3PL model for equating observed scores below the chance level. One obvious procedure to obtain the relation between the tests below the level of the pseudo-chance level is to use (linear) interpolation. Lord (1980) suggested an alternative, a raw-score adaptation of the IRT-equating method. In this procedure, the distribution of θ is estimated for some group. Given this distribution, the marginal distributions of x and y can be estimated. Next, X and Y can be equated through equipercentile equating. The outcome depends to some extent on the group.

11.7.4 Other models

The equating methodology can be extended to the linking of tests with polytomous items. Cohen and Kim (1998) present an overview of linking methods under the graded response model. This model sometimes is used in connection with the judgment by raters of constructed responses. The fact that judges play a role complicates the linking process. For, it is by no means sure that judges have, for example, a stable year-to-year severity of judgment (Ercika et al., 1998; Tate, 1999).

11.8 Concluding remarks

At a certain moment, a new test form Y, equated with an old test form X, is replaced on its turn by a more recent test form Z. So, we are not ready after equating the two test forms. After some time, we obtained a chain of multiple test forms equated to each other. With more than two test forms, we can use alternative equating designs. In Figure 11.4, the three possible designs with three different tests X, Y, and Z are displayed. For more information on flexible equating designs, see Exhibit 11.3.

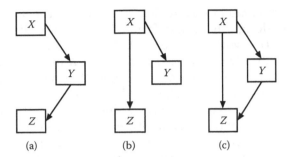

Figure 11.4 The three equating designs with three test forms X, Y, and Z.

Exhibit 11.3   A history of equating

The Test of English as a Foreign Language (TOEFL) has had a history of a frequent introduction of new test forms. These tests should yield scores that can be used exchangeably. Equating has been used to eliminate possible differences between the test forms. First, conventional equating methods were used. From 1978 equating tests has been based on IRT, using the three-parameter logistic model.

Cowell (1982) describes the history of this change in practice. He also notes the change that IRT equating makes possible. With IRT equating, data can be obtained from different previous tests. This allows us to try out items in pretests given to relatively small groups of persons. This means an improvement in test security.

With IRT equating, it is possible to have a new test that entirely consists of previously administered items with item characteristics estimated on a common scale—that is, items selected from an item bank. This "pre-equating" of items fastens the process of scoring.

If tests can be made from an existing pool of items, adaptive testing comes into reach, a fact also mentioned by Cowell. Two sections of the computer-based version of the TOEFL are adaptive.

One alternative is to equate test form Y with test form X using a common subtest of Y and X, and next test form Z with Y using a common subtest of Z and Y (Figure 11.4a). Test Z has no items in common with X, but test Z and test X are on the same scale as well through test Y. A weakness of the design is the risk that the inevitable equating errors cumulate. That risk is avoided in the design of Figure 11.4b. In this design, both test forms Y and Z are directly equated with form X. Herewith, test form X has become a standard. This solution has a disadvantage, too. Items from X may become obsolete. Items from X might also become known, so that many examinees can learn these items by heart. The third design (Figure 11.4c) is more balanced. In this design, test form Z has common items with both forms X and Y. We may verify whether direct equating of test forms Z and X results in the same score transformation as equating forms Z and X through test Y. So, this design has a control mechanism. Another advantage of this design in comparison with the design in Figure 11.4b is that a limited number of items from test form X is needed in later test forms.

When IRT is used, an adequate choice of the model is very important. If guessing occurs or is highly likely, then we prefer to use a model with a pseudo-chance-level parameter, otherwise equating errors are bound to be made, especially in the low score range and with vertical equating.

In this chapter, we took the position that equating is necessary. However, whether one should equate test forms is something that has to be decided. In the final exhibit, Exhibit 11.4, this problem is discussed.

Exhibit 11.4   To equate or not to equate

How do we decide whether or not to equate multiple test forms? So far, no systematic study answering this question has been performed. It is clear that the equating of test scores is a prerequisite for large programs of educational and psychological testing where multiple test forms are involved of, for example, scholastic aptitude. The criteria to decide whether or not to equate forms depend at least on the following objectives:

Choosing the optimal design for data on the forms to be equated
Selecting the best equating procedure that meets the conditions or desirable properties for equating
Minimizing statistical errors when conducting an equating study

These objectives, however, are intertwined. Minimizing statistical errors, systematic as well as random, depends on the study design and the equating procedure used. In design considerations, sample size plays a crucial role and sample size affects statistical error. To be sure, minimizing equating error is a major goal when deciding whether or not to equate forms, when designing equating studies, and when conducting equating.

No conclusive answer on the question of this exhibit can be given. Kolen (1999, pp. 171–174) has a number of relevant comments to make on the equating methodology to date. He also briefly describes some testing programs with equating studies. Of course, much can be learned from the relevant research reports on testing programs where equating is a must, or at least an essential standard.

Exercises

11.1 We have two tests forms X and Y. In a large random selection of persons, the mean score on Y is 60.0 and the standard deviation is 16.0. In a second random selection, the mean score on X is 55.0 and the standard deviation is 17.0. We want to equate Y with X. With which score on X does the score $y = 50$ correspond according to the linear equating method?

11.2 We have two groups of persons. One group is administered test form Y, the other is administered test form X. Suppose we know that the group that answered test form Y is somewhat better than the other group. What can you say about the x score equivalent to $y = 50$ in Exercise 11.1? Explain your answer.

11.3 V is a subtest of test X. Assume that test X is the sum of subtest V and $k - 1$ tests parallel to subtest V. Prove that k ($= \sigma_{T(X)}/\sigma_{T(V)}) = \sigma_X^2/\sigma_{XV}$ (Equation 11.9).

11.4 We have two test forms, X and Y, each with five items. Both tests are analyzed with the Rasch model. The items are numbered consecutively in the table below. We notice that two items, items 3 and 5, are common to X and Y.

Item	1	2	3	4	5	6	7	8
b test Y	−1.5	−1.0	0.0	0.5	2.0			
b test X			−0.5		1.5	−0.5	−0.5	0.0

Compute the estimated item parameters of the items of Y on the scale defined by test form X. Give the relationship between true scores for X and Y for $\theta = -4.0$ (0.5) 4.0.

11.5 In a study, test X has been administered to a group of high-ability examinees, and test Y to a low-ability group. Both groups also have been tested with test V. In a second study, all examinees have been tested with test V. The examinees with relatively high scores on V have been tested with test X, the other examinees have been tested with test Y. All three tests are supposed to measure the same characteristic. Test X is relatively difficult, and test Y relatively easy, but we do not know how much the tests differ in difficulty level. What is the characteristic difference between the two studies and what are the consequences of this difference?

Answers

1.1 Researcher A probably will obtain a lower gain for the best pupils than will researcher B. On the raw score scale, the score cannot exceed the maximum score. In the study by researcher A there is a ceiling effect.

1.2 Imagine what might happen to the ranks if a player was added. The score scale is a typical example of an ordinal scale.

2.1 The test center compares the persons with other persons who have been tested at the same moment of the day. For the test center, a fixed moment is part of the definition of the true score. The persons who are tested may have another point of view. They might compare their results with those of other persons who were tested by other centers at different moments. In that case, one should generalize accidental variation in test administration times when defining true score. If the outcome of a test is notably influenced by the moment of the day on which the test is administered, it is relevant to know how the test developer dealt with moment of the day in the norming study.

2.2 It is possible that persons get used to the test format or that fatigue plays a role when the persons take the second test. In that case, the condition of experimental independence is not satisfied. In a study with two tests X and Y, the role of fatigue can be controlled to a certain extent by using a test design in which one group of persons is administered test X first and test Y next, and another group of persons is administered the tests in reverse order.

3.1 Application of Formula 3.3 gives 5.0.

3.2 The result of the computations is presented in the table below. The reliability increases strongly at first. For larger values of k, the increase becomes smaller. For large values of k, the reliability approaches the limiting value of one.

k	1	2	3	4	5	6	7	8	9	10	11	12	13	14
$\rho_{xx'}$	0.50	0.67	0.75	0.80	0.83	0.86	0.88	0.89	0.90	0.91	0.92	0.92	0.93	0.93

3.3 The standard error of estimation equals the standard error of measurement times the square root of the reliability coefficient. For $\rho_{xx'} = 0.5$, the square root equals 0.71; for $\rho_{xx'} = 0.9$, it equals 0.95. So, for $\rho_{xx'} = 0.5$, the ratio of standard errors equals 0.71; for $\rho_{xx'} = 0.9$, the ratio equals 0.95. The Kelley estimate of true score is equal to 35.0 for $\rho_{xx'} = 0.5$, and equal to 31.0 for $\rho_{xx'} = 0.9$. For low reliability, confidence intervals for the true score based on the observed score and the standard error of measurement deviate strongly from confidence intervals obtained using the Kelley point estimate of true score and the standard error of estimation. For high reliabilities, the difference between the two approaches is relatively small.

3.4 Use the formula for the correction for attenuation. From this formula, it can be deduced that

$$\rho_{XY} \leq \sqrt{\rho_{XX'}}\sqrt{\rho_{YY'}} \leq \sqrt{\rho_{XX'}}$$

Clearly, the correlation with a criterion cannot exceed the square root of the reliability. In other words, the correlation of the observed scores with their true scores gives an upper limit to the correlation of a measurement instrument with other variables. The maximum correlation for reliability equal to 0.49 is 0.7.

3.5

$$\rho_{X(k)Y} = \frac{\mathrm{cov}(X(k),Y)}{\sigma_Y \sigma_{X(k)}} = \frac{\mathrm{cov}\left[\left(kT_X + \sum E_{X(i)}\right), Y\right]}{\sigma_Y \sqrt{k^2 \sigma_{T_X}^2 + k \sigma_{E_X}^2}}$$

$$= \frac{k\,\mathrm{cov}(T_X, Y)}{\sigma_Y \sqrt{k^2 \rho_{XX'} \sigma_X^2 + k(1 - \rho_{XX'}) \sigma_X^2}} = \frac{k\rho_{XY}}{\sqrt{k^2 \rho_{XX'} + k(1 - \rho_{XX'})}}$$

When k goes to infinity, the formula can be simplified to

$$\rho_{X(\infty)Y} = \frac{\rho_{XY}}{\sqrt{\rho_{XX'}}}$$

(assuming $\rho_{XX'} > 0$).

4.1 The variance of total scores—the sum of entries in the table—equals 89. The sum of the variances of the subtests equals 37. Coefficient α is equal to 0.88.

4.2 In order to be able to apply Formula 4.7, we have to determine the values of the factor loadings a_1, a_2, and a_3. The factor loading a_1 equals the square root of $(\sigma_{12}\sigma_{13})/\sigma_{23}$ (see Formula 4.6). The factor loading a_2 equals the square root of $(\sigma_{21}\sigma_{23})/\sigma_{13}$. The factor loading a_3 equals the square root of $(\sigma_{31}\sigma_{32})/\sigma_{12}$. The computation of these factor loadings results in $a_1 = 2.0$, $a_2 = 3.0$, and $a_3 = 4.0$. The reliability according to Formula 4.7 is $9.0^2/89.0 = 0.91$, a value that is a bit higher than the reliability estimated with coefficient α.

4.3

$$\alpha = \frac{n^2 \text{ave(cov)}}{\sigma_X^2} = \frac{n^2 \text{ave(cov)}}{n \text{ave}\left(\sigma_i^2\right) + n(n-1)\text{ave(cov)}}$$

$$= \frac{n^2 r^*}{n + n(n-1)r^*} = \frac{n\rho}{1+(n-1)\rho}$$

where ave stands for $average$; $r^* = \text{ave(cov)}/\text{ave}(\sigma_i^2)$. Because the items are parallel, r^* is equal to the common value of the inter-item correlation ρ. The result is identical to the Spearman-Brown formula.

4.4 The correlation between X_i ($i = 1, 2$) and an arbitrary third test Y is given by

$$\rho_{X_iY} = \frac{\text{cov}(X_i,Y)}{\sigma_{X_i}\sigma_Y} = \frac{a_i}{\sigma_{X_i}}\frac{\text{cov(T,}Y)}{\sigma_Y} = \sqrt{\rho_{X_iX_i'}}\frac{\text{cov(T,}Y)}{\sigma_Y}$$

where T is the true score on the common true-score scale defined by $\mu_T = 0.0$ and $\sigma_T^2 = 1.0$. So, congeneric measurements have identical patterns of correlations with other variables.

4.5 a. The observed score variance σ_D^2 is equal to $\sigma_X^2 + \sigma_Y^2 - 2\sigma_X\sigma_Y\rho_{XY}$. This gives us the value 9.6 for the observed variance of the difference scores. The true-score variance $\sigma_{T(D)}^2$ is equal to $\sigma_{T(X)}^2 + \sigma_{T(Y)}^2 - 2\sigma_{T(X)}\sigma_{T(Y)}\rho_{T(X)T(Y)}$. The true-score variances are obtained by multiplying the observed-score variances with the reliabilities. The covariance of the true differences, $\sigma_{T(X)}\sigma_{T(Y)}\rho_{T(X)T(Y)}$, is equal to the covariance of the observed differences, $\sigma_X\sigma_Y\rho_{XY}$. This gives us the value 3.2 for $\sigma_{T(D)}^2$. The reliability of the difference scores is low: $3.2/9.6 = 0.33$. The low value is due to the high correlation between the true scores of the two tests.

   b. The variance of the differences is 9.6. This is much larger than 6.4, the value of the error variance obtained from Equation 4.18. The variance of the differences is larger because the true scores on X and Y differ.

4.6 The condition of experimental independence might be violated. This can affect the reliability estimate as well as reliability. The violation of experimental independence can be eliminated through a redefinition of the items. Items belonging together might be treated as a single item when reliability is estimated from responses to the items. This redefinition of the item level might produce a new problem. The true-score variance of an item consisting of many subitems can be much larger than the true-score variances of other items. In this case, α would underestimate the reliability. The effect of large differences between true-score variances can be avoided by grouping all items in item clusters before reliability is estimated.

4.7 The inequality of means and correlations indicates that the tests are not parallel or tau-equivalent. The equal covariances indicate equal true-score variances. So, the three tests are essentially tau-equivalent.

4.8 When no specific effect like a learning effect is expected, we may assume that the true scores on both occasions are equal. The expected observed score on both occasions equals the true score. The expected true score on the second occasion given an observed score equal to 30 at the first occasion is 35.0, under the assumption that the regression of true score on observed score is linear (application of the Kelley formula). The expected difference score equals 5 (35 − 30). This is the regression effect.

4.9 The true-score variance of the composite is $0.8 \times 25.0 + 0.6 \times 25.0$. The observed-score variance of the composite is $25.0 + 25.0$. The reliability is 0.7; see also Equation 4.11. For the reliability of the lengthened test, we use the Spearman–Brown formula. We obtain 0.824 as the reliability of the lengthened test. We see that the reliability of a test composed of noncorrelating subtests can be high. Of course, it makes no sense to combine noncorrelating subtests into one test.

5.1 In the computation of the variance, we divide the numerator by the number of persons minus one. The item variances are 0.1342, 0.2211, 0.2395, 0.1974, 0.2211, 0.2605, 0.2632, 0.2395, 0.2605, and 0.2632. The sum of the variances equals 2.300. The variance of the total scores is 4.011. Coefficient α is $(10/9)(1 - 2.300/4.011) = 0.47$.

The results of the analysis of variance are given in the following table:

Source of Variation	Sum of Squares	Degrees of Freedom	Mean Squares
Persons	7.620	19	0.4011
Items	2.920	9	0.3244
Residual	36.080	171	0.2110
Total	46.620		

The variance component for persons equals $(0.4011 - 0.2110)/10 = 0.019$. The variance component for items equals 0.006. The residual (0.211) is by far the largest component, more than ten times as large as the variance component for persons. Measurement errors as well as the interaction between persons and items are part of this component. Due to the fact that there are no replications, the error and interaction components cannot be separated. The interaction component must be larger than zero because the items, which differ in difficulty level, cannot be essential tau-equivalent measurements of the underlying trait.

The generalizability coefficient for a test of ten items is equal to 0.47. This value is the same as the value of coefficient α (as it should be—the coefficients are mathematically identical).

The example was chosen only to keep the computational burden low. It should be clear that the estimated variance components are unreliable due to the small number of persons and items.

5.2 The estimated residual component equals 0.65. The variance component for the interaction items $\times$ judges σ_{ij}^2 is $(45.65 -0.65)/500 = 0.090$. The (estimated) variance component σ_{pj}^2 is equal to 0.010, σ_{pi}^2 equals 1.500, σ_j^2 equals 0.050, σ_i^2 equals

0.50, and σ_p^2 equals 0.175. The residual component is the largest component. The variance components involving judges are relatively small: judges seem to be reasonably well exchangeable. The generalizability coefficient for 15 items and 4 judges equals 0.61 (Formula 5.11 or 5.12).

5.3　See Formula 5.14.

　　a. $n'_i = 30$; $n'_j = 4$; the estimate of $E\rho^2$ is 0.75.

　　b. $n'_i = 60$; $n'_j = 2$; the estimate of $E\rho^2$ is 0.83.

　　In (a) as well as in (b) the total number of observations per person is twice the number used in the generalizability study. An increase of the number of items has a strong effect on generalizability, even if the number of judges decreases. One might have expected this result in view of the outcome of Exercise 5.2. The variance components in which judges are involved are relatively small; the judges are relatively well exchangeable.

5.4　The number of observations for a particular combination of p and i is n_j. This is the coefficient of the variance component for the interaction of p and i: $a = c = e = n_j$. The other coefficients are $b = n_i n_j$ and $d = n_p n_j$.

5.5　When the correlation between judges is computed, the test items are regarded as fixed. The correlation can be written as the generalizability coefficient:

$$E\rho_{\text{Rel}}^2 = \frac{\left[\sigma_p^2 + \sigma_{pi}^2/n_i\right]}{\left[\sigma_p^2 + \sigma_{pi}^2/n_i\right] + \left(\left[\sigma_{pj}^2 + \sigma_{pij}^2/n_i\right] + \sigma_e^2/n_i\right)/n_j}$$

with n_j equal to 1.

5.6　The relative error variance is the error variance that plays a role in the generalizability coefficient for the crossed $p \times i \times j$ design. The error variance is equal to

$$\sigma_{\text{Rel}}^2 = \sigma_{pi}^2/n_i + \sigma_{pj}^2/n_j + \sigma_{pij,e}^2/n_i n_j$$

For the absolute error variance, the variance components in which p is not involved are also relevant. The absolute error variance is

$$\sigma_{\text{Abs}}^2 = \sigma_i^2/n_i + \sigma_j^2/n_j + \sigma_{ij}^2/n_i n_j + \sigma_{pi}^2/n_i + \sigma_{pj}^2/n_j + \sigma_{pij,e}^2/n_i n_j$$

5.7　In this exercise, we have a design in which persons are nested within judges. The variance of the mean score for a judge, for random samples of 50 persons, equals 2.0 (the person variance within judges divided by sample size, i.e., the number of persons). If there are no differences between judges, a variance between judges equal to 2.0 is expected. The variance between judges is higher: 9.0. Obviously judges are not equally lenient. One might consider the possibility to correct psychometrically for the differences in leniency. A full correction for the obtained mean differences between judges does not seem appropriate because some of the differences might be due to differences between the random samples of persons. The reliability of the effects of the judges equals 7.0 (observed variance for judges minus error variance) divided by 9.0 (observed variance). Application of Kelley's formula gives an estimated effect for judge 1 equal to $(7.0/9.0) \times (32.0 - 35.0) = -2.33$. The effect for judge 2 is estimated as 0.0, and the estimated effect for judge 3 is 2.33. Judge 1 is too harsh. One might correct for this by adding 2.33 to all judgments by this judge.

The analysis is not fully satisfactory. The estimation of the true variation between judges is based on just three judges. The compromise between no correction for judges and a full correction is, however, attractive. One might use the outcome not only for the purpose of statistically correcting scores, but also to stimulate the reorientation of the training of judges and the reformulation of judgmental instructions.

6.1 In this exercise, the binomial model is to be used with parameters $\zeta = 0.8$ and $n = 10$. The probabilities of 8, 9, and 10 correct responses have to be summed. The probability of 8 correct is 0.3020, the probability of 9 correct 0.2684, and the probability of 10 correct 0.1074. The probability of 8 or more items correct equals 0.678.

6.2 a. The proportion of correct responses for item 8 is 0.65, the item–rest correlation is 0.543. The item-test regression is given in the table below. For this particular item with a high item–rest correlation, the proportion of correct responses as a function of total score increases strongly in the score range 3 to 7.

Total Score	0	1	2	3	4	5	6	7	8	9	10
Proportion Correct	—	—	0.0	—	0.33	0.25	0.50	1.0	1.0	1.0	—

b. Item 6 has the lowest item–rest correlation. The correlation is unsatisfactory (it is negative!). So, item 6 should be eliminated first. Also, when item 6 is dropped, the increase in coefficient alpha is highest.

6.3 We can estimate the reliability with KR21. The average proportion correct equals 0.75, and the test variance equals 2.25. The resulting estimate equals 0.185.

6.4 Here we have an application of Equation 6.11.

$$P(x_1 = 1, x_2 = 1 \mid \zeta) = 0.56$$
$$P(x_1 = 1, x_2 = 0 \mid \zeta) = 0.14$$
$$P(x_1 = 0, x_2 = 1 \mid \zeta) = 0.24$$
$$P(x_1 = 0, x_2 = 0 \mid \zeta) = 0.06$$

6.5 Keats' solution assumes that the variance of item means given true score is relatively high in the middle score range. We can compute the item-test regressions like in Figure 6.1. When many item-test regressions cross each other in the middle score range, this assumption is untenable.

6.6 The error variance is $0.6 \times 0.4 + 0.7 \times 0.3 + 0.8 \times 0.2 = 0.61$.

The true proportion correct ζ_p is equal to 0.7. The binomial error variance is slightly higher: $3 \times 0.7 \times (1 - 0.7) = 0.63$.

The variance of the item difficulties at $\zeta = \zeta_p$ is $(0.1^2 + 0.0^2 + 0.1^2)/3 = 0.02/3$. The difference between the binomial error variance and the variance in the generalized binomial model is equal to $3 \times 0.02/3$.

6.7 The covariance between the item and the total test is $(n - 1)$ cov $+ s^2$, where n is the number of items, cov is the covariance between the items, and s^2 is the variance of the item. The variance of the test scores is $ns^2 + n(n - 1)$cov. The item-total correlation can be computed from the covariance and the variances. The item-rest correlation can be computed using Equation 6.20. The results are given in the following table.

n	r_{it}	r_{ir}
10	0.529	0.372
20	0.490	0.406
40	0.469	0.426

The r_{it} is spuriously high. The effect of the item as part of the total test strongly diminishes as test length increases. The value of r_{it} decreases. The value r_{ir} depends only on the reliability of the rest-test. The reliability increases with test length; so, r_{ir} increases with test length. The difference between the two indexes of item discrimination power decreases with increasing test length.

7.1 The fact that persons have been selected does not imply that the variance diminishes. In the exercise we have an example of selection which results in a higher variance in the selected group. Application of Formula 7.3 gives the value of 0.66 for the correlation in the total group.

7.2 We compare the relative frequencies of A and B for each score level. In order to obtain these relative frequencies, we multiply the frequencies of group A by four.

Score	0	1	2	3	4	5	6	7	8	9	10
$4 \times f_A$	0.172	0.436	0.520	0.696	0.868	0.696	0.348	0.172	0.088	0.0	0.0
f_B	0.0	0.0	0.0	0.0	0.045	0.091	0.136	0.182	0.227	0.182	0.136

From this table the posterior probabilities can be obtained. For score level 7, for example, the posterior probability for A equals 0.172/(0.172 + 0.182), and the posterior probability for B equals 0.182/(0.172 + 0.182). From score 7 onward, the (posterior) probability of dealing with a B person is higher than the probability of dealing with an A person. From score 7 through score 10 we therefore classify a person as belonging to group B. With persons who belong to B we make a wrong classification in 27.2% of the cases (all B persons with a score lower than 7). With persons belonging to population A we make a mistake in only 6.5% of the cases (all A persons with a score equal to or higher than 7). In 1 out of the 5 cases a B person is involved and then we make a wrong classification in 27.2% of the cases, in 4 out of the 5 cases we have an A person and then we make a wrong classification in 6.5% of the cases (see the frequency distribution f_A). On average, we make a mistake in $100 \times (0.2 \times 0.272 + 0.8 \times 0.065) = 10.6\%$ of the cases. The relatively large error with respect to population B is due to the fact that this population is so much smaller than population A.

7.3 The posterior probability of belonging to group B increases for every score level, as might be inferred from Figure 7.2 and Equation 7.4. Therefore, the critical score for allocation to B instead of A moves to the left. The data in the table of Exercise 7.2 are relevant for a base rate equal to 0.5. We notice that the probability of belonging to B exceeds the probability of belonging to A at a score equal to or higher than 6. Persons with a score equal to or higher than 6 can be classified as belonging to group B. Many more persons are classified as B persons.

7.4 The optimal cut score is the score for which the expected true score equals the criterion of mastery, 0.70. The expected true score for a given score x is given by Kelley's formula. With the given criterion, mean score, and reliability, we find that the value for the optimal cut score is 55.0 (from $70.0 = 0.25x + (1 - 0.25) \times 75$), well below the criterion on the true score scale. This is due to the low test reliability.

7.5 The proportion of correct classifications is 0.80. The proportion of correct classifications expected by chance is $0.7^2 + 0.3^2 = 0.58$. The value of κ is 0.524.

8.1 The eigenvalues are $\lambda_1 = 4.818$, $\lambda_2 = 1.658$, and $\lambda_3 = 0.746$. The first two components are the only eigenvalues larger than 1. The first two components account for $100 \times (4.4818 + 1.658)/10 = 64.8\%$ of the total variance; the third factor increases the percentage of variance accounted for with only 7.5%; and the fourth factor adds another 6.1%. These two findings support the decision to keep two components.

8.2 The eighth test, numerical puzzles, has the smallest communality: 0.356 ($0.5137^2 + 0.3042^2$). Its distance to the origin in Figure 8.2 is $h = 0.597$.

8.3 There are two latent variables, F_1 and F_2; it is arbitrary which one is called the first.

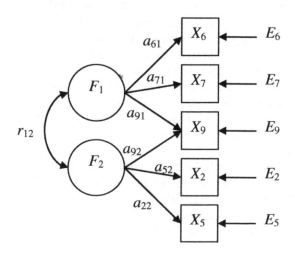

9.1 The answers to Exercise 9.1 are given in the table below.

θ	−2.0	−1.5	−1.0	−0.5	0.0	0.5	1.0	1.5	2.0
$P(\theta)$	0.12	0.18	0.27	0.38	0.50	0.62	0.73	0.82	0.88

9.2 In the Rasch model, one restriction is needed in order to fix the latent scale. We take the restriction $b_1 = 0$. Next, we estimate θ_1 using the logarithm of $P_1(\theta_1)/[1 - P_1(\theta_1)]$. This logarithm equals $\theta_1 - b_1 = \theta_1$. The value of θ_1 is −0.5. In a similar way, we obtain θ_2; θ_2 equals 0.5.

Knowing θ_1, we can compute the item parameter of the second item, $b_2 = -\ln\{P_2(\theta_1)/[1 - P_2(\theta_1)]\} + \theta_1$. The value of b_2 is −0.25. The value of b_2 can also be obtained from the equation $b_2 = -\ln\{P_2(\theta_2)/[1 - P_2(\theta_2)]\} - \theta_2$. Using this equation, we obtain −0.40 as the value of b_2. The second computation of the item parameter is not in agreement with the first computation. This means that the Rasch model cannot describe the probabilities.

9.3 The values of the likelihood for the different levels of θ from the exercise are given in the fifth column in the table below.

θ	$P_1(\theta)$	$P_2(\theta)$	$Q_3(\theta) =$ $1 - P_3(\theta)$	$P(\mathbf{x}\|\theta) = P_1(\theta)$ $P_2(\theta)Q_3(\theta)$	$\theta P(\mathbf{x}\|\theta)g(\theta)/$ $P(\mathbf{x})$
−1.0	0.3775	0.2689	0.8176	0.0830	−0.0957
−0.5	0.5000	0.3775	0.7311	0.1380	−0.0795
0.0	0.6225	0.5000	0.6225	0.1937	0.0
0.5	0.7311	0.6225	0.5000	0.2275	0.1311
1.0	0.8176	0.7311	0.3775	0.2257	0.2600

$P(\mathbf{x}) = 0.8679 \times 0.20$ EAP = 0.2159

a. The maximum value of the likelihood $P(\mathbf{x}|\theta)$ is obtained in the table for $\theta = 0.5$. The ML estimate of θ must lie in the neighborhood of this value of θ (i.e., between 0.0 and 1.0). For a value of θ slightly above 0.5, the likelihood exceeds the likelihood at $\theta = 0.5$. (The derivative of the log likelihood [Equation 9.39] is positive at $\theta = 0.5$.) So, the ML estimate lies in the interval 0.5 to 1.0.

b. Due to the equality of the latent classes, the computation of $P(\mathbf{x})$ can be simplified. The value of the EAP is 0.22.

This estimate is closer to the population mean than the ML estimate.

9.4 The information values (see Equation 9.33) for the items are given in the rightmost column of the following table.

Item	$p(\theta)$	$P(\theta)$	$p(\theta)[1 - p(\theta)]$	$I(\theta)$
1	0.378	0.378	0.235	0.235
2	0.269	0.269	0.197	0.786
3	0.269	0.452	0.197	0.351

9.5 The true variance equals 0.300. The error variance for each level of θ is obtained by taking $1/I(\theta)$. The average error variance equals 0.108. The reliability (true variance divided by the sum of the true variance and the average error variance) equals 0.735.

10.1 The question itself is biased. It is possible that the second item is biased against the focal group, but it is equally possible that item 1 is biased against the reference group. There is not enough evidence to choose between these two rival hypotheses and the third hypothesis of no bias. The difference between the outcomes of the two items might also be due to a higher discrimination of item 2.

10.2 We are looking for a two-item test for which the minimum of $I(\theta_1)$ and $I(\theta_2)$ is maximal. The test information equals the sum of the item informations. The item informations for the two levels of θ are as in the table below.

b_i	$I_i(\theta_1)$	$I_i(\theta_2)$
−0.50	0.2500	0.1966
−0.30	0.2475	0.2139
0.0	0.2350	0.2350
0.25	0.2179	0.2461
0.50	0.1966	0.2500

For the combination of items 2 and 4, $\min\{I(\theta_1), I(\theta_2)\}$ equals 0.46, obtained at θ_2. All other item combinations have a smaller value for the minimum of $I(\theta_1)$ and $I(\theta_2)$. The combination of items 2 and 4 is optimal.

10.3 The next item to be presented is the item with the highest item information at $\theta = 0.20$. The item informations for this value of θ are 0.22, 0.82, 0.12, 0.25, and 0.53. The second item has the highest item information (0.82). Therefore, item 2 is the item to be presented next.

11.1 From Formulas 11.2 through 11.4, we obtain $y' = 1.0625y - 8.75$. The score y = 50 is equivalent to x score 44.375.

11.2 If the group that made test Y is a bit better, then the average score on this test is somewhat too high in comparison to the average x score. The factor B from Formula 11.4 is too low, as well as the equivalent score on X, obtained in this way. The equivalent score x should be higher than 44.375, the value obtained in Exercise 11.1. The correction that is needed can be provided by information on an anchor test V. The average score v in the group persons to which test Y was administered, is higher than the average score v in the group to which test X was administered. This results in a higher value for the equivalent score x in Formula 11.8.

11.3 In order to obtain test X, subtest V must be lengthened by a factor k. The observed-score variance of test X can be written as

(a) $$\sigma_X^2 = k^2\sigma_T^2 + k\sigma_E^2$$

where σ_T^2 is the true-score variance of subtest V and σ_E^2 the error variance of subtest V. Let V be the first of the k subtests. The covariance between test X and subtest V is

(b) $$\sigma_{XV} = \text{cov}\left(kT + \sum_{i=1}^{k} E_i, T + E_1\right) = k\sigma_T^2 + \sigma_E^2$$

If we divide the result of (a) by the result of (b), we obtain factor k.

11.4 The average b of the common items is equal to 1.0 for test Y and 0.5 for test X. In order to bring the b parameters from test Y to the scale of test X, the following transformation is to be applied: $b_Y{}^* = b_Y - 1.0$ (average parameter value of item 3 and 5 in Y) + 0.5 (average parameter value of item 3 and 5 in X): $b_Y{}^* = b_Y - 0.5$.

The item parameters of the items from test Y on the scale of X are –2.0, –1.5, –0.5, 0.0, and 1.5. The true scores on tests X and Y for a given value of θ are computed by means of Equation 6.17.

The relation between the true score for $\theta = -4.0$ (0.5) 4.0 is as shown in the table.

θ	τ_Y	τ_X
−4.0	0.25	0.11
−3.5	0.39	0.18
−3.0	0.59	0.29
−2.5	0.86	0.45
−2.0	1.21	0.70
−1.5	1.62	1.04
−1.0	2.08	1.48
−0.5	2.55	2.00
0.0	3.00	2.55
0.5	3.43	3.08
1.0	3.80	3.56
1.5	4.12	3.96
2.0	4.38	4.28
2.5	4.58	4.51
3.0	4.72	4.68
3.5	4.82	4.80
4.0	4.89	4.87

For all true scores on Y, the corresponding true scores on X are lower. Test form Y is the easier test form.

11.5 In the first study, the two examinee groups are defined on the basis of an external criterion. In the second study (the study with two-stage testing), the group definition is based on the common routing test V. Due to the fact that test V is not perfectly reliable, we have a regression effect on the test given as the second test. Study 1 presents the design of a vertical equating study with common or anchor test V. In study 1, unbiased item parameter estimates for all three tests can be obtained. This is not the case in study 2 unless the regression effect is effectively dealt with. Application of the mean and sigma method would not result in correct estimates on a common scale.

References

Adams, R. J., Wilson, M., and Wang, W. C. (1997). The multinomial logit model. *Applied Psychological Measurement, 21*, 1–23.

Akkermans, W. (2000). Modelling sequentially scored item responses. *British Journal of Mathematical and Statistical Psychology, 53*, 83–98.

Albert, J. H. (1992). Bayesian estimation of normal ogive item response curves using Gibbs sampling. *Journal of Educational Statistics, 17*, 251–269.

Alf, E. F., and Dorfman, D. D. (1967). The classification of individuals into two criterion groups on the basis of a discontinuous payoff function. *Psychometrika, 32*, 115–123.

Allalouf, A., Hambleton, R. K., and Sireci, S. G. (1999). Identifying the causes of DIF in translated verbal items. *Journal of Educational Measurement, 36*, 185–198.

American Council on Education. (1995*). Guidelines for Computer-Adaptive Test Development and Use in Education*. Washington, DC: Author.

American Educational Research Association. (1955). *Technical Recommendations for Achievement Tests*. Washington, DC: Author.

American Psychological Association. (1954). Technical recommendations for psychological tests and diagnostic techniques. *Psychological Bulletin, 51* (Suppl.).

American Psychological Association. (1966). *Standards for Educational and Psychological Tests and Manuals*. Washington, DC: Author.

American Psychological Association, American Educational Research Association, and National Council on Measurement in Education. (1974). *Standards for Educational and Psychological Tests*. Washington, DC: American Psychological Association.

American Psychological Association, American Educational Research Association, and National Council on Measurement in Education. (1985). *Standards for Educational and Psychological Testing*. Washington, DC: American Psychological Association.

American Psychological Association, American Educational Research Association, and National Council on Measurement in Education. (1999). *Standards for Educational and Psychological Testing*. Washington, DC: American Psychological Association.

Anastasi, A. (1954). *Psychological Testing*. New York: MacMillan.

Andersen, E. B. (1972). The numerical solution of a set of conditional estimation equations. *Journal of the Royal Statistical Society, Series B, 34,* 42–54.

Andersen, E. B. (1973). A goodness of fit test for the Rasch-model. *Psychometrika, 38,* 123–140.

Andersen, E. B. (1977). Sufficient statistics and latent trait models. *Psychometrika, 42,* 69–81.

Andersen, E. B. (1983). A general latent structure model for contingency data. In H. Wainer and S. Messick (Eds.), *Principals of Modern Psychological Measurement* (pp. 117–138). Hillsdale, NJ: Lawrence Erlbaum Associates.

Anderson, N. H. (1961). Scales and statistics: parametric and nonparametric. *Psychological Bulletin, 58,* 305–316.

Andrich, D. (1978). A rating formulation for ordered response categories. *Psychometrika, 43,* 561–573.

Andrich, D. (1999). Rating scale analysis. In J. P. Keeves and G. N. Masters (Eds.), *Advances in Measurement in Educational Research and Assessment* (pp. 110–121). Amsterdam: Pergamon.

Andrich, D., Sheridan, B., and Luo, G. (2000). *RUMM2010: A Windows Interactive Program for Analyzing Data with Rasch Unidimensional Models for Measurement,* RUMM Laboratory, Perth, Western Australia.

Angoff, W. H. (1953). Test reliability and effective test length. *Psychometrika, 18,* 1–14.

Angoff, W. H. (1971). Scales, norms and equivalent scores. In R. L. Thorndike (Ed.), *Educational Measurement* (2nd ed., pp. 509–600). Washington, DC: American Council on Education.

Armor, D. J. (1974). Theta reliability and factor scaling. In H. L. Costner (Ed.), *Sociological Methodology 1973–1974* (pp. 17–50). San Francisco: Jossey-Bass.

Assessment Systems Corporation. (1996). *User's Manual for the XCALIBRE Marginal Maximum-Likelihood Estimation Program.* St. Paul, MN: Assessment Systems Corp.

Attali, Y. (2005). Reliability of speeded number-right multiple-choice tests. *Applied Psychological Measurement, 29,* 357–368.

Baker, F. B., and Kim, S. -H. (2004). *Item Response Theory: Parameter Estimation Techniques.* New York: Dekker.

Belov, D. I., and Armstrong, R. D. (2005). Monte Carlo test assembly for item pool analysis. *Applied Psychological Measurement, 29,* 239–261.

Bickel, P., Buyske, S., Chang, H., and Ying, Z. (2001). On maximizing item information and matching difficulty with ability. *Psychometrika, 66,* 69–77.

Birnbaum, A. (1968). Some latent trait models and their use in inferring an examinee's ability. In F. M. Lord and M. R. Novick, *Statistical Theories of Mental Test Scores.* Reading, MA: Addison-Wesley.

Blinkhorn, S. F. (1997). Past imperfect, future conditional: fifty years of test theory. *British Journal of Mathematical and Statistical Psychology, 50,* 175–185.

Blok, H. (1985). Estimating the reliability, validity and invalidity of essay ratings. *Journal of Educational Measurement, 22,* 41–52.

Bock, R. D. (1972). Estimating item parameters and latent ability when responses are scored in two or more nominal categories. *Psychometrika, 37,* 29–51.

Bock, R. D., and Aitkin, M. (1981). Marginal maximum likelihood estimation of item parameters: application of an EM algorithm. *Psychometrika, 46,* 443–459; *47,* 369 (Errata).

Bock, R. D., Gibbons, R., and Muraki, E. (1988). Full-information factor analysis. *Applied Psychological Measurement, 13,* 261–280.

Bock, R. D. and Mislevy, R. J. (1982). Adaptive EAP estimation of ability in a microcomputer environment. *Applied Psychological Measurement, 6,* 431–444.

Borsboom, D., Mellenbergh, G. J., and Van Heerden, J. (2004). The concept of validity. *Psychological Review, 111,* 1061–1071.

Bouwmeester, S., and Sijtsma, K. (2004). Measuring the ability of transitive reasoning, using product and strategy information. *Psychometrika, 69,* 123–146.

Braun, H. I. (1988). Understanding scoring reliability: experiments in calibrating essay readers. *Journal of Educational Statistics, 13,* 1–18.

Brennan, R. L. (1987). Problems, perspectives, and practical issues in equating. *Applied Psychological Measurement, 11,* 221–306.

Brennan, R. L. (1998). Raw-score conditional standard errors of measurement in generalizability theory. *Applied Psychological Measurement, 22,* 307–331.

Brennan, R.L. (2001). *Generalizability Theory.* New York: Springer.

Brennan, R. L., and Kane, M. T. (1977). An index of dependability for mastery tests. *Journal of Educational Measurement, 14,* 277–289.

Bryant, F. B., and Yarnold, P. R. (1995). Principal-components analysis and exploratory and confirmatory factor analysis. In L. G. Grimm and P. R. Yarnold (Eds.), *Reading and Understanding Multivariate Statistics* (pp. 99–136). Washington DC: American Psychological Association.

Burr, J. A., and Neselroade, J. R. (1990). Change measurement. In A. von Eye (Ed.), *Statistical Methods in Longitudinal Research* (Vol. 1, pp. 3–34). Boston: Academic Press.

Byrne, B. M. (1998). *Structural Equation Modeling with LISREL, PRELIS and SIMPLIS: Basic Concepts, Applications and Programming.* Hillsdale, NJ: Lawrence Erlbaum Associates.

Byrne, B. M. (2001). *Structural Equation Modeling with AMOS: Basic Concepts, Applications and Programming.* Mahwah, NJ: Lawrence Erlbaum Associates.

Byrne, B. M. (2006). *Structural Equation Modeling with EQS: Basic Concepts, Applications and Programming* (2nd ed.). Mahwah, NJ: Lawrence Erlbaum Associates.

Camilli, G., and Shepard, L. A. (1994). *Methods for Identifying Biased Test Items*. Thousand Oaks, CA: Sage.

Campbell, D. T., and Fiske, D. W. (1959). Convergent and divergent validation by the multitrait–multimethod matrix. *Psychological Bulletin, 56,* 81–105.

Carlson, J. F. (1998). Review of the Beck Depression Inventory (1993 Revised). In J. C. Impara and B. S. Plake (Eds.), *The Thirteenth Mental Measurement Yearbook* (pp. 117–119). Lincoln, NE: The Buros Institute of Mental Measurements.

Chang, Y. -C. I. (2005). Application of sequential interval estimation to adaptive mastery testing. *Psychometrika, 70,* 685–713.

Cizek, G. J. (1999). *Cheating on Tests: How to Do It, Detect It, and Prevent It.* Mahwah, NJ: Lawrence Erlbaum Associates.

Cohen, A. S., and Kim, S. -H. (1998). An investigation of linking methods under the graded response model. *Applied Psychological Measurement, 22,* 116–130.

Cohen, J. (1960). A coefficient of agreement for nominal scales. *Educational and Psychological Measurement, 20,* 37–46.

Cohen, L. (1979). Approximate expressions for parameter estimates in the Rasch-model. *British Journal of Mathematical and Statistical Psychology, 32,* 113–120.

Cole, N. S., and Moss, P. A. (1989). Bias in test use. In R. L. Linn (Ed.), *Educational Measurement* (3rd ed., pp. 201–219). New York: American Council on Education.

Cook, L. L., and Petersen, N. S. (1987). Problems related to the use of conventional and item response theory equating methods in less than optimal circumstances. *Applied Psychological Measurement, 11,* 225–244.

Cook, T. D., Campbell, D. T., and Peracchio, L. (1990). Quasi experimentation. In M. D. Dunnette and L. M. Hough (Eds.), *Handbook of Industrial and Organizational Psychology* (2nd ed., pp. 491–576). Palo Alto, CA: Consulting Psychologists Press.

Cook, T. D., and Shadish, W. R. (1994). Social experiments and some developments over the past fifteen years. *Annual Review of Psychology, 45,* 545–580.

Coombs, C. H. (1964). *A Theory of Data.* New York: Wiley.

Cooper, H., and Hedges, L. V. (Eds.) (1994). *The Handbook of Research Synthesis.* New York: Russell Sage Foundation.

Cowell, W. R. (1982). Item-response-theory pre-equating in the TOEFL testing program. In P.W. Holland and D. B. Rubin (Eds.), *Test Equating* (pp. 149–161). New York: Academic Press.

Crano, W. D. (2000). The multitrait–multimethod matrix as synopsis and recapitulation of Campbell's views on the proper conduct of social inquiry. In L. Bickman (Ed.), *Research Design: Donald Campbell's Legacy* (Vol. 2, pp. 37–61). Thousand Oaks, CA: Sage.

Cronbach, L. J. (1951). Coefficient alpha and the internal structure of tests. *Psychometrika, 16*, 297–334.

Cronbach, L. J. (1971). Test validation. In R. L. Thorndike (Ed.), *Educational Measurement* (2nd ed., pp. 443–507). Washington, DC: American Council on Education.

Cronbach, L. J. (1990). *Essentials of Psychological Testing* (5th ed.). New York: Harper & Row.

Cronbach, L. J., and Furby, L. (1970). How should we measure "change" or should we? *Psychological Bulletin, 74*, 68–80.

Cronbach, L. J., and Gleser, G. C. (1965). *Psychological Tests and Personnel Decisions* (2nd ed.). Urbana: University of Illinois Press.

Cronbach, L. J., Gleser, G. C., Nanda, H., and Rajaratnam, N. (1972). *The Dependability of Behavioral Measurements: Theory of Generalizability for Scores and Profiles*. New York: Wiley.

Cronbach, L. J., Linn, R. L., Brennan, R. L., and Haertel, E. H. (1997). Generalizability analysis for performance assessment of student achievement or school effectiveness. *Educational and Psychological Measurement, 57*, 373–399.

Cronbach, L. J., and Meehl, P. E. (1955). Construct validity in psychological tests. *Psychological Bulletin, 52*, 281–302.

Cronbach, L. J., Rajaratnam, N., and Gleser, G. C. (1963). Theory of generalizability: a liberalization of reliability theory. *British Journal of Statistical Psychology, 16*, 137–163.

Cronbach, L. J., and Warrington, W. G. (1952). Efficiency of multiple-choice tests as a function of spread of item difficulties. *Psychometrika, 17*, 127–147.

Croon, M. (1991). Investigating Mokken scalability of dichotomous items by means of ordinal latent class analysis. *British Journal of Mathematical and Statistical Psychology, 44*, 315–331.

Daniel, M. H. (1999). Behind the scenes: using new measurement methods on the DAS and KAIT. In S. E. Embretson and S. L. Herschberger (Eds.), *The New Rules of Measurement: What Every Psychologist and Educator Should Know* (pp. 37–63). Mahwah, NJ: Lawrence Erlbaum Associates.

De Gruijter, D. N. M. (1988). Standard errors of item parameter estimates in incomplete designs. *Applied Psychological Measurement, 12*, 109–116.

De Gruijter, D. N. M., and Van der Kamp, L. J. Th. (1991). Generalizability theory. In R. K. Hambleton and J.N. Zaal (Eds.), *Advances in Educational and Psychological Testing* (pp. 45–68). Boston: Kluwer.

De la Torre, J., and Patz, R. J. (2005). Making the most of what we have: a practical application of multidimensional item response theory in test scoring. *Journal of Educational and Behavioral Statistics, 30*, 295–311.

De Leeuw, J. and Verhelst, N. (1986). Maximum likelihood estimation in generalized Rasch models. *Journal of Educational Statistics, 11*, 183–196.

Dickens, W. T., and Flynn, J. R. (2001). Heritability estimates versus large environmental effects: the IQ paradox resolved. *Psychological Review, 108*, 346–369.

Donoghue, R. R., and Isham, P. (1998). A comparison of procedures to detect item parameter drift. *Applied Psychological Measurement, 22*, 33–51.

Dorans, N. J. (1990). Equating methods and sampling designs. *Applied Measurement in Education, 3*, 1–17.

Dorans, N. J., and Holland, P. W. (1993). DIF detection and description: Mantel-Haenszel and standardization. In P. W. Holland and H. Wainer (Eds.), *Differential Item Functioning* (pp. 35–66). Hillsdale, NJ: Lawrence Erlbaum Associates.

Dorans, N. J., and Kulick, E. M. (1986). Demonstrating the utility of the standardization approach to assessing unexpected differential item performance on the Scholastic Aptitude Test. *Journal of Educational Measurement, 23*, 355–368.

Drasgow, F., Levine, M. V., and McLaughlin, M. E. (1987). Detecting inappropriate test scores with optimal and practical appropriateness indices. *Applied Psychological Measurement, 11*, 59–79.

Drasgow, F., Levine, M. V., and Williams, E. A. (1985). Appropriateness measurement with polychotomous item response models and standardized indices. *British Journal of Mathematical and Statistical Psychology, 38*, 67–86.

Ebel, R. L. (1961). Must all tests be valid? *American Psychologist, 16*, 640–647.

Efron, B., and Tibshirani, R. J. (1993). *An Introduction to the Bootstrap.* New York: Chapman and Hall.

Embretson, S. E. (1984). A general latent trait model for response processes. *Psychometrika, 49*, 175–186.

Embretson, S. E. (1991). A multidimensional latent trait model for measuring learning and change. *Psychometrika, 56*, 495–516.

Embretson, S. E., and Prenovost, L. K. (1999). Item response theory in assessment research. In P. C. Kendall, J. N. Butcher, and G. N. Holmbeck (Eds.), *Handbook of Research Methods in Clinical Psychology* (pp. 276–294). New York: Wiley.

Embretson, S. E., and Reise, S. P. (2000). *Item Response Theory for Psychologists.* Mahwah, NJ: Lawrence Erlbaum Associates.

Ercika, K., Schwartz, R. D., Julian, M. W., Burket, G. R., Weber, M. M., and Link, V. (1998). Calibration and scoring of tests with multiple-choice and constructed-response item types. *Journal of Educational Measurement, 35*, 137–154.

Falmagne, J. -C. (1989). A latent trait theory via a stochastic learning theory for a knowledge space. *Psychometrika, 54*, 283–303.

Feldt, L. S. (1984). Some relationships between the binomial error model and classical test theory. *Educational and Psychological Measurement, 44*, 883–891.

Feldt, L. S., and Brennan, R. L. S. (1989). Reliability. In R. L. Linn (Ed.), *Educational Measurement* (3rd ed., pp. 105–146). New York: American Council on Education.

Feldt, L. S., and Qualls, A. L. (1996). Estimation of measurement error variance at specific score levels. *Journal of Educational Measurement, 33*, 141–156.

Feldt, L. S., Steffen, M., and Gupta, N. C. (1985). A comparison of five methods for estimating the standard error of measurement at specific score levels. *Applied Psychological Measurement, 9*, 351–361.

Fhanér, S. (1974). Item sampling and decision making in achievement testing. *British Journal of Mathematical and Statistical Psychology, 27*, 172–175.

Finch, H. (2005). The mimic model as a method for detecting DIF: comparison with Mantel-Haenszel, SIBTEST, and the IRT likelihood ratio, *Applied Psychological Measurement, 29*, 278–295.

Fischer, G. H. (1976). Some probabilistic models for measuring change. In D. N. M. de Gruijter and L. J. Th. van der Kamp (Eds.), *Advances in Psychological and Educational Measurement* (pp. 97–110). New York: Wiley.

Fischer, G. H. (1983). Logistic latent trait models with linear constraints. *Psychometrika, 48*, 3–26.

Fleiss, J. L., Cohen, J., and Everitt, B. S. (1969). Large-sample standard errors of kappa and weighted kappa. *Psychological Bulletin, 72*, 323–327.

Flynn, J. R. (1987). Massive gains in 14 nations: what IQ tests really measure. *Psychological Bulletin, 101*, 171–191.

Fraser, C. (1988). *NOHARM: A Computer Program for Fitting both Unidimensional and Multidimensional Normal Ogive Models of Latent Trait Theory*. Armidale, NSW: University of New England.

Furneaux, W. D. (1960). Intellectual abilities and problem-solving behaviour. In H. J. Eysenck (Ed.), *Handbook of Abnormal Psychology* (pp. 167–192). London: Pitman.

Gelman, A., Carlin, J. B., Stern, H. S., and Rubin, D. B. (2004). *Bayesian Data Analysis*. Boca Raton, FL: Chapman and Hall.

Gessaroli, M. E., and De Champlain, A. F. (2005). Test dimensionality: Assessment of. In B. S. Everitt and D. C. Howell (Eds.), *Encyclopedia of Statistics in Behavioral Science* (Vol. 4, pp. 2014–2021). Chichester: Wiley.

Gifi, A. (1990). *Nonlinear Multivariate Analysis*. Chichester: Wiley.

Glas, C. A. W., and Falcón, J. C. S. (2003). A comparison of item-fit statistics for the three-parameter logistic model. *Applied Psychological Measurement, 27*, 87–106.

Glas, C. A. W., and Meijer, R. R. (2003). A Bayesian approach to person-fit analysis in item response theory models. *Applied Psychological Measurement, 27*, 217–233.

Goldstein, H., and Wood, R. (1989). Five decades of item response modelling. *British Journal of Mathematical and Statistical Psychology, 42*, 139–167.

Gorsuch, L. R. (1983). *Factor Analysis* (2nd ed.). Hillsdale, NJ: Lawrence Erlbaum Associates.

Gulliksen, H. (1950). *Theory of Mental Tests*. New York: Wiley.

Guttman, L. (1945). A basis for test-retest reliability. *Psychometrika, 10*, 255–282.

Guttman, L. (1950). The basis for scalogram analysis. In S. A. Stouffer, L. Guttman, E. A. Suchman, P. F. Lazarsfeld, S. A. Star, and J. A. Clausen (Eds.), *Measurement and Prediction* (Vol. IV, pp. 60–90). Princeton, NJ: Princeton University Press.

Guttman, L. (1953). A special review of Harold Gulliksen, Theory of mental tests, *Psychometrika, 18*, 123–130.

Haebara, T. (1980). Equating logistic ability scales by weighted least squares method. *Japanese Psychological Research, 22*, 144–149.

Haladyna, T. M. (1999). *Developing and Validating Multiple-Choice Test Items* (2nd ed.). Mahwah, NJ: Lawrence Erlbaum Associates.

Hambleton, R. K., and Novick, M. R. (1973). Toward an integration of theory and method for criterion-referenced tests. *Journal of Educational Measurement, 10*, 159–170.

Hambleton, R. K., and Pitoniak, M. J. (2006). Setting performance standards. In R. L. Brennan (Ed.), *Educational Measurement* (4th ed., pp. 433–470). Westport, CT: American Council on Education/Praeger.

Hambleton, R. K., Swaminathan, H., and Rogers, H. J. (1991). *Fundamentals of Item Response Theory*. Newbury Park, CA: Sage.

Hand, D. J. (1997). *Construction and Assessment of Classification Rules*. New York: Wiley.

Hanson, B. A., Zeng, L., and Kolen, M. J. (1993). Standard errors of Levine linear equating. *Applied Psychological Measurement, 17*, 225–237.

Heinen, T. (1996). *Latent Class and Discrete Latent Trait Models. Similarities and Differences*. Thousand Oaks, CA: Sage.

Henrysson, S. (1963). Correction of item-total correlations in item analysis. *Psychometrika, 28*, 211–218.

Holland, P. W. (1990). On the sampling theory foundations of item response theory models. *Psychometrika, 55*, 577–601.

Holland, P. W., and Wightman, L. E. (1982). Section pre-equating: a preliminary investigation. In P. W. Holland and D. B. Rubin (Eds.), *Test Equating* (pp. 271–297). New York: Academic Press.

Holzinger, K. J., and Swineford, F. (1939). A study in factor analysis: the stability of a bi-factor solution. *Supplementary Educational Monographs, No 48*. Chicago: Department of Education, University of Chicago.

Hoyt, C. J. (1941). Test reliability estimated by analysis of variance. *Psychometrika, 6*, 153–160.

Huitzing, H. A., Veldkamp, B. P., and Verschoor, A. J. (2005). Infeasibility in automated test assembly models: a comparison study of different methods. *Journal of Educational Measurement, 42*, 223–243.

Hunter, J. E., and Schmidt, F. L. (1990). *Methods of Meta-Analysis*. Newbury Park: Sage.

Huynh, H. (1978). Reliability of multiple classifications. *Psychometrika, 43*, 317–325.

Huynh, H. (1994). On the equivalence between a partial credit model and a set of independent Rasch binary items. *Psychometrika, 59*, 111–119.

Jackson, P. H. (1973). The estimation of true score variance and error variance in the classical test theory model. *Psychometrika, 38*, 183–201.

Jansen, P. G. W., and Roskam, E. E. (1986). Latent trait models and dichotomization of graded responses. *Psychometrika, 51*, 69–91.

Jarjoura, D. (1983). Best linear prediction of composite universe scores. *Psychometrika, 48*, 525–539.

Jarjoura, D. (1986). An estimator of examinee-level measurement error variance that considers test form difficulty adjustments. *Applied Psychological Measurement, 10*, 175–186.

Jarjoura, D., and Brennan, R. L. (1982). A variance components model for measurement procedures associated with a table of specifications. *Applied Psychological Measurement, 6*, 161–171.

Jolliffe, I. T. (2002). *Principal Component Analysis* (2nd ed.). New York: Springer.

Jöreskog, K. G. (1971). Statistical analysis of sets of congeneric tests. *Psychometrika, 36*, 109–133.

Jöreskog, K. G., and Sörbom, D. (1993). *LISREL 8 User's Reference Guide.* Chicago: Scientific Software International.

Judd, C. M., and McClelland, G. H. (1998). Measurement. In D. T. Gilbert, S. T. Fiske, and G. Lindzey (Eds.), *The Handbook of Social Psychology* (4th ed., Vol. 1, pp. 180–232). Boston, MA: McGraw-Hill.

Kane, M. T. (2006). Validation. In R. L. Brennan (Ed.), *Educational Measurement* (4th ed., pp. 17–64). Westport, CT: American Council on Education/Praeger.

Keats, J. A. (1957). Estimation of error variances of test scores. *Psychometrika, 22*, 29–41.

Kelderman, H. (1984). Loglinear Rasch model tests. *Psychometrika, 49*, 223–245.

Kelley, T. L. (1947). *Fundamentals of Statistics.* Cambridge: Harvard University Press.

Kendall, M. G., and Stuart, A. (1961). *The Advanced Theory of Statistics* (Vol. II). London: Griffin.

Kenny, D. A., and Kashy, D. A. (1992). The analysis of the multitrait–multimethod matrix using confirmatory factor analysis. *Psychological Bulletin, 112*, 165–172.

Kim, S. -H. (2001). An evaluation of a Markov chain Monte Carlo method for the Rasch model. *Applied Psychological Measurement, 25*, 163–176.

Kim, S. -H., and Cohen, A. S. (1998). Detection of differential item functioning under the graded response model with the likelihood ratio test. *Applied Psychological Measurement, 22*, 345–355.

Kish, L. (1987). *Statistical Design for Research.* Wiley: New York.

Kok, F. G., Mellenbergh, G. J., and Van der Flier, H. (1985). Detecting experimentally induced item bias using the iterative logit method. *Journal of Educational Measurement, 22*, 295–303.

Kolen, M. J. (1999). Equating of tests. In G. M. Masters and J. P. Keeves (Eds.), *Advances in Measurement in Educational Research and Assessment* (pp. 164–175). Amsterdam: Pergamon.

Kolen, M. J., and Brennan, R. L. (1995). *Test Equating: Methods and Practices.* New York: Springer-Verlag.

Kuder, G. F., and Richardson, M. W. (1937). The theory of the estimation of test reliability. *Psychometrika, 2*, 151–160.

Lane, S., and Stone, C. A. (2006). Performance assessment. In R. L. Brennan (Ed.), *Educational Measurement* (4th ed., pp. 387–431). Westport, CT: American Council on Education/Praeger.

Lee, W. -C., Brennan, R. L., and Kolen, M. J. (2000). Estimators of conditional scale-score standard errors of measurement: a simulation study. *Journal of Educational Measurement, 37*, 1–20.

Lee, W. -C., Hanson, B. A., and Brennan, R. L. (2002). Estimating consistency and accuracy indices for multiple classifications. *Applied Psychological Measurement, 26*, 412–432.

Li, Y., Bolt, D. M., and Fu, J. (2006). A comparison of alternative models for testlets. *Applied Psychological Measurement, 30*, 3–21.

Li, Y. H., and Schafer, W. D. (2005). Trait parameter recovery using multidimensional computerized adaptive testing in reading and mathematics. *Applied Psychological Measurement, 29*, 3–25.

Lindley, D. V. (1971). The estimation of many parameters. In V. P. Godambe and D. A. Sprott (Eds.), *Foundation of Statistical Inference* (pp. 435–447). Toronto: Holt, Rinehart and Winston.

Linn, R. L., and Harnisch, D. L. (1981). Interactions between item content and group membership on achievement test items. *Journal of Educational Measurement, 18*, 109–118.

Linn, R. L., Levine, M. V., Hastings, C. N., and Wardrop, J. L. (1981). Item bias in a test of reading comprehension. *Applied Psychological Measurement, 5*, 159–173.

Little, R. J. A., and Rubin, D. B. (1987). *Statistical Analysis with Missing Data.* New York: Wiley.

Longford, N. T. (1994). Reliability of essay rating and score adjustment. *Journal of Educational and Behavioral Statistics, 19*, 171–200.

Lord, F. M. (1952). A theory of test scores. *Psychometric Monograph*, No. 7. Chicago: University of Chicago Press.

Lord, F. M. (1953). On the statistical treatment of football numbers. *American Psychologist, 8*, 750–751.

Lord, F. M. (1954). Further comment on "football numbers." *The American Psychologist, 9*, 264–265.

Lord, F. M. (1955). Equating test scores—a maximum likelihood solution. *Psychometrika, 20*, 193–200.

Lord, F. M. (1958). Some relations between Guttman's principal components of scale analysis and other psychometric theory. *Psychometrika, 23*, 291–296.

Lord, F. M. (1962). Cutting scores and errors of measurement. *Psychometrika, 27*, 19–30.

Lord, F. M. (1977). Optimal number of choices per item—a comparison of four approaches. *Journal of Educational Measurement, 14*, 33–38.

Lord, F. M. (1980). *Applications of Item Response Theory to Practical Testing Problems.* Hillsdale, NJ: Lawrence Erlbaum Associates.

Lord, F. M., and Novick, M. R. (1968). *Statistical Theories of Mental Test Scores.* Reading, MA: Addison-Wesley.

Marcoulides, G. A. (1999). Generalizability theory: picking up where the Rasch IRT model leaves off? In S. E. Embretson and S. L. Hershberger (Eds.), *The New Rules of Measurement: What Every Psychologist and Educator Should Know* (pp. 129–152). Mahwah, NJ: Lawrence Erlbaum Associates.

Masters, G. N. (1982). A Rasch-model for partial credit scoring. *Psychometrika, 47*, 149–174.

Masters, G. N. (1999). Partial credit model. In J. P. Keeves and G. N. Masters (Eds.), *Advances in Measurement in Educational Research and Assessment* (pp. 98–109). Amsterdam: Pergamon.

Maxwell, A. E., and Pilliner, A. E. G. (1968). Deriving coefficients of reliability and agreement for ratings. *British Journal of Mathematical and Statistical Psychology, 21*, 105–116.

McDonald, R. P. (1968). A unified treatment of the weighting problem. *Psychometrika, 33*, 351–381.

McDonald, R. P. (1997). Normal-ogive multidimensional model. In W. J. van der Linden and R. K. Hambleton (Eds.), *Handbook of Modern Item Response Theory* (pp. 258–269). New York: Springer-Verlag.

McDonald, R. P. (1999). *Test Theory: A Unified Treatment.* Mahwah, NJ: Lawrence Erlbaum Associates.

McKinley, R. L., and Mills, C. N. (1985). A comparison of several goodness-of-fit statistics. *Applied Psychological Measurement, 9*, 49–57.

Meijer, R. R., and Sijtsma, K. (2001). Methodology review: evaluating person fit. *Applied Psychological Measurement, 25*, 107–135.

Meredith, W., and Kearns, J. (1973). Empirical Bayes point estimates of latent trait scores without knowledge of the trait distribution. *Psychometrika, 38*, 533–554.

Messick, S. (1989). Validity. In R. L. Linn (Ed.), *Educational Measurement* (3rd ed., pp. 13–103). New York: American Council on Education.

Messick, S. (1994). The interplay of evidence and consequences in the validation of performance assessments. *Educational Researcher, 23*, 13–23.

Messick, S. (1995). Validity of psychological assessment: validation of inferences from persons' responses and performances as scientific inquiry into score meaning. *American Psychologist, 50,* 741–749.

Michell, J. (1999*). Measurement in Psychology: Critical History of a Methodological Concept.* Cambridge: Cambridge University Press.

Michell, J. (2005). Measurement: overview. In B. S. Everitt and D. C. Howell (Eds.), *Encyclopedia of Statistics in Behavioral Science* (Vol. 3, pp. 1176–1183). Chichester: Wiley.

Millsap, R. E., and Everson, H. T. (1993). Methodology review: statistical approaches for assessing measurement bias. *Applied Psychological Measurement, 17,* 297–334.

Mislevy, R. J. (1983). Item response models for grouped data. *Journal of Educational Statistics, 8,* 271–288.

Mislevy, R. J., and Bock, R. D. (1990). BILOG 3: item analysis and test scoring with binary logistic models. Mooresville, IN: Scientific Software.

Mokken, R. J. (1971). *A Theory and Procedure of Scale Analysis with Applications in Political Research.* New York; Berlin: Walter de Gruyter (Mouton).

Mokken, R. J., Lewis, C., and Sijtsma, K. (1986). Rejoinder to "The Mokken scale: a critical discussion." *Applied Psychological Measurement, 10,* 279–285.

Molenaar, I. W. (1983). Some improved diagnostics for failure of the Rasch-model. *Psychometrika, 48,* 49–72.

Molenaar, I. W. (1997). Nonparametric models for polytomous responses. In W. J. van der Linden and R. K. Hambleton (Eds.), *Handbook of Modern Item Response Theory* (pp. 369–380). New York: Springer-Verlag.

Molenaar, I. W., and Hoijtink, H. (1990). The many null distributions of person-fit indices. *Psychometrika, 55,* 75–106.

Molenaar, I. W., and Sijtsma, K. (2002*). Introduction to Nonparametric Item Response Theory.* Thousand Oaks, CA: Sage.

Mosier, C. I. (1947). A critical examination of the concept of face validity. *Educational and Psychological Measurement, 7,* 191–205.

Muraki, E. (1990). Fitting a polytomous item response model to Likert-type data. *Applied Psychological Measurement, 14,* 59–71.

Muraki, E., and Bock, R. D. (1997). *PARSCALE. IRT Based Test Scoring and Item Analysis for Graded Open-Ended Exercises and Performance Tasks.* Chicago: Scientific Software.

Muraki, E., and Carlson, J. E. (1995). Full-information factor analysis for polytomous item responses. *Applied Psychological Measurement, 19,* 73–90.

Muthén, B. O. (1984). A general structural equation model with dichotomous, ordered categorical and continuous latent variable indicators. *Psychometrika, 49,* 115–132.

Muthén, B. O. (2002). Beyond *SEM*: general latent variable modeling. *Behaviormetrika, 29,* 81–117.

Nandakumar, R., Yu, F., Li, H. -H., and Stout, W. (1998). Assessing unidimensionality of polytomous data. *Applied Psychological Measurement, 22,* 99–115.

Nishisato, S. (1994). *Elements of Dual Scaling: An Introduction to Practical Data Analysis.* Hillsdale, NJ: Lawrence Erlbaum Associates.

Nishisato, S. (1980). *Analysis of Categorical Data: Dual Scaling and Its Applications*. Toronto: University of Toronto Press.

Novick, M. R., and Jackson, P. H. (1974). *Statistical Methods for Educational and Psychological Research*. New York: McGraw-Hill.

Oud, J. H. L., Van den Bercken, J. H., and Essers, R. J. (1990). Longitudinal factor score estimation using the Kalman filter. *Applied Psychological Measurement, 14*, 395–418.

Overall, J. E. (1965). Reliability of composite ratings. *Educational and Psychological Measurement, 25*, 1011–1022.

Pandey, T. N., and Hubert, L. (1975). An empirical comparison of several interval estimation procedures for coefficient alpha. *Psychometrika, 40*, 169–181.

Panter, A. T., Kimberly, A. S., and Dahlstrom, W. G. (1997). Factor analytic approaches to personality item-level data. *Journal of Personality Assessment, 68*, 561–589.

Patz, R. J., and Junker, B. W. (1999). A straightforward approach to Markov Chain Monte Carlo methods for item response models. *Journal of Educational and Behavioral Statistics, 24*, 146–178.

Patz, R. J., Junker, B. W., Johnson, M. S., and Marino, L. T. (2002). The hierarchical rater model for rated test items and its application to large-scale educational assessment data. *Journal of Educational and Behavioral Statistics, 27*, 341–384.

Pearson, E. S., and Hartley, H. O. (1970). *Biometrika Tables for Statisticians* (3rd ed., Vol. I). Cambridge: Cambridge University Press.

Petersen, N. S., Cook, L. L., and Stocking, M. L. (1983). IRT versus conventional equating methods: a comparative study of scale stability. *Journal of Educational Statistics, 8*, 137–156.

Petersen, N. S., Kolen, M. J., and Hoover, H. D. (1989). Scaling, norming, and equating. In R. L. Linn (Ed.), *Educational Measurement* (3rd ed., pp. 221–262). New York: American Council on Education.

Petersen, N. S., and Novick, M. R. (1976). An evaluation of some models for culture-fair selection. *Journal of Educational Measurement, 13*, 3–29.

Popham, W. J., and Husek, T. R. (1969). Implications of criterion-referenced measurement. *Journal of Educational Measurement, 6*, 1–9.

Rae, G. (2006). Correcting coefficient alpha for correlated errors: is α_K a lower bound to reliability? *Applied Psychological Measurement, 30*, 56–59.

Rajaratnam, N. (1960). Reliability formulas for independent decision data when reliability data are matched. *Psychometrika, 25*, 261–271.

Rajaratnam, N., Cronbach, L. J., and Gleser, G. C. (1965). Generalizability of stratified parallel tests. *Psychometrika, 30*, 39–56.

Ramsay, J. O., and Silverman, B. W. (2002). *Applied Functional Data Analysis: Methods and Case Studies*. New York: Springer.

Rao, C. R., and Sinharay, S. (Eds.). (2007). *Handbook of statistics*, Vol. 26 (*Psychometrics*). Amsterdam: Elsevier.

Rasch, G. (1960). *Probabilistic Models for Some Intelligence and Attainment Tests*. Copenhagen: Danish Institute for Educational Research.

Raykov, T. (1998). A method for obtaining standard errors and confidence intervals of composite reliability for congeneric items. *Applied Psychological Measurement, 22*, 369–374.

Raykov, T. (2001). Bias of coefficient for fixed congeneric measures with correlated errors. *Applied Psychological Measurement, 25*, 69–76.

Reckase, M. D. (1997). The past and future of multidimensional item response theory. *Applied Psychological Measurement, 21*, 25–36.

Reise, S. P. (1999). Personality measurement; issues viewed through the eyes of IRT. In S. E. Embretson and S. L. Herschberger (Eds.), *The New Rules of Measurement: What Every Psychologist and Educator Should Know* (pp. 219–241). Mahwah, NJ: Lawrence Erlbaum Associates.

Reise, S. P., and Waller, N. G. (1990). Fitting the two-parameter model to personality data. *Applied Psychological Measurement, 14*, 45–58.

Reise, S. P., Widaman, K. F., and Pugh, R. H. (1993). Confirmatory factor analysis and item response theory: two approaches for exploring measurement invariance. *Psychological Bulletin, 114*, 552–566.

Revuelta, J., and Ponsoda, V. (1998). A comparison of item exposure control methods in computerized adaptive testing. *Journal of Educational Measurement, 35*, 311–327.

Rijmen, F., De Boeck, P., and Van der Maas, H. L. J. (2005). An IRT model with a parameter-driven process for change. *Psychometrika, 70*, 651–669.

Rogers, W. T., and Harley, D. (1999). An empirical comparison of three- and four-choice items and tests: susceptibility to testwiseness and internal consistency reliability. *Educational and Psychological Measurement, 59*, 234–247.

Rogosa, D., Brandt, D., and Zimowski, M. (1982). A growth curve approach to the measurement of change. *Psychological Bulletin, 90*, 726–748.

Rogosa, D., and Ghandour, G. (1991). Statistical models for behavioral observations. *Journal of Educational Statistics, 16*, 157–252.

Roskam, E. E. (1997). Models for speed and time-limit tests. In W. J. van der Linden and R. K. Hambleton (Eds.), *Handbook of Modern Item Response Theory* (pp. 187–208). New York: Springer-Verlag.

Rossi, N., Wang, X., and Ramsay, J. O. (2002). Nonparametric item response function estimates with the EM algorithm. *Journal of Educational and Behavioral Statistics, 27*, 291–317.

Rost, J. (1990). Rasch models in latent classes: an integration of two approaches to item analysis. *Applied Psychological Measurement, 14*, 271–282.

Rost, J. (1991). A logistic mixture distribution model for polychotomous item responses. *British Journal of Mathematical and Statistical Psychology, 44*, 75–92.

Roussos, L. A., Stout, W. F., and Marden, J. I. (1998). Using new proximity measures with hierarchical cluster analysis to detect multidimensionality. *Journal of Educational Measurement, 35*, 1–30.

Rudner, L. M. (1983). Individual assessment accuracy. *Journal of Educational Measurement, 20,* 207–219.

Rudner, L. M., Getson, P. R., and Knight, D. L. (1980). Biased item detection techniques. *Journal of Educational Statistics, 5,* 213–233.

Sackett, P. R., and Yang, H. (2000). Correction for range restriction: an expanded typology. *Journal of Applied Psychology, 85,* 112–118.

Samejima, F. (1969). Estimation of latent ability using a response pattern of graded scores. *Psychometric Monograph, No. 18.* Iowa City, IA: Psychometric Society.

Samejima, F. (1973). A comment on Birnbaum's three-parameter logistic model in the latent trait theory. *Psychometrika, 38,* 221–233.

Samejima, F. (1979). A new family of models for the multiple-choice item. *Research Report 79-4.* University of Tennessee, Knoxville, TN.

Sands, W. A., Waters, B. K., and McBride, J. R. (Eds.). (1997). *Computerized Adaptive Testing: From Inquiry to Operation.* Washington, DC: American Psychological Association.

Scheuneman, J. D., and Bleistein, C. A. (1999). Item bias. In G. N. Masters and J. P. Keeves (Eds.), *Advances in Measurement in Educational Research and Assessment* (pp. 220–234). Amsterdam: Pergamon.

Schmitt, N., Coyle, B. W., and Saari, B. B. (1977). A review and critique of analyses of multitrait–multimethod matrices. *Mutivariate Behavioral Research, 12,* 447–478.

Schmitt N., and Stults, D. M. (1986). Methodology review: analysis of multitrait–multimethod matrices. *Applied Psychological Measurement, 10,* 1–22.

Shadish, W. R., Cook, T. D., and Campbell, D. T. (2002). *Experimental and Quasi-Experimental Designs for Generalized Causal Inference.* Boston, MA: Houghton-Mifflin.†

Shavelson, R. J., and Webb, N. M. (1991). *Generalizability Theory: A Primer.* Newbury Park, CA: Sage; with corrections: 5th print 1995.

Shavelson, R. J., Webb, N. M., and Rowley, G. L. (1989). Generalizability theory. *American Psychologist, 44,* 922–932.

Shealy, R. T., and Stout, W. F. (1993). A model-based standardization approach that separates true bias/DIF from group ability differences and detects bias/DTF as well as item bias/DIF. *Psychometrika, 58,* 159–194.

Shepard, L. A., Camilli, G., and Williams, D. M. (1985). Validity of approximation techniques for detecting item bias. *Journal of Educational Measurement, 22,* 77–105.

Shi, J. Q., and Lee, S. Y. (1997). A Bayesian estimation of factor score in confirmatory factor model with polytomous, censored or truncated data. *Psychometrika, 62,* 29–50.

Sinharay, S., Johnson, M. S., and Stern, H. S. (2006). Posterior predictive assessment of item response theory models. *Applied Psychological Measurement, 30,* 298–321.

Sirotnik, K., and Wellington, R. (1977). Incidence sampling: an integrated theory for "matrix sampling." *Journal of Educational Measurement, 14*, 343–399.

Skaggs, S. G., and Lissitz, R. W. (1986). IRT equating: relevant issues and a review of recent research. *Review of Educational Research, 56*, 495–529.

Snijders, T. A. B., and Bosker, R. J. (1999). *Multilevel Analysis: An Introduction to Basic and Advanced Multilevel Modeling.* London: Sage.

Sotaridona, L., Van der Linden, W. J., and Meijer, R. R. (2006). Detecting answer copying using the kappa statistic. *Applied Psychological Measurement, 30*, 412–431.

Spiegelhalter, D., Thomas, A., Best, N., and Lunn, D. (2003). *WinBUGS User Manual, Version 1.4.* Cambridge, UK: University of Cambridge, Institute of Public Health, MRC Biostatistics Unit (www.mrc-bsu.cam.ac.uk/bugs).

Stevens, S. S. (1951). Mathematics, measurement, and psychophysics. In S.S. Stevens (Ed.), *Handbook of Experimental Psychology* (pp. 1–49). New York: Wiley.

Stevens, S. S. (1968). Measurement, statistics, and the schemapiric view. *Science, 161*, 849–856.

Stocking, M. L., and Lord, F. M. (1983). Developing a common metric in item response theory. *Applied Psychological Measurement, 7*, 201–210.

Stout, W. (1987). A nonparametric approach for assessing latent trait unidimensionality. *Psychometrika, 52*, 589–617.

Stout, W. (2002). Psychometrics: from practice to theory and back. *Psychometrika, 67*, 485–518.

Subkoviak, M. J. (1984). Estimating the reliability of mastery–nonmastery classifications. In R. A. Berk (Ed.), *A Guide to Criterion-Referenced Test Construction* (pp. 267–291). Baltimore: Johns Hopkins University Press.

Suen, H. K., and Ary, D. (1989). *Analyzing Quantitative Behavioral Observation Data.* Hillsdale, NJ: Lawrence Erlbaum Associates.

Swaminathan, H., and Gifford, J. A. (1986). Bayesian estimation in the three-parameter logistic model. *Psychometrika, 51*, 589–601.

Tate, R. L. (1999). A cautionary note on IRT-based linking of tests with polytomous items. *Journal of Educational Measurement, 36*, 336–346.

Taylor, H. C., and Russell, J. T. (1939). The relationship of validity coefficients to the practical effectiveness of tests in selection: discussion and tables. *Journal of Applied Psychology, 23*, 565–578.

Tellegen, A. (1982). *A Brief Manual for the Multidimensional Personality Questionnaire.* Unpublished manuscript, University of Minnesota.

Ten Berge, J. M. F., and Sŏcan, G. (2004). The greatest lower bound to the reliability of a test and the hypothesis of unidimensionality. *Psychometrika, 69*, 613–625.

Theunissen, T. J. J. M. (1985). Binary programming and test design. *Psychometrika, 50*, 411–420.

Thissen, D. (1991). *MULTILOG User's Guide: Multiple Categorical Item Analysis and Test Scoring Using Item Response Theory.* Chicago: Scientific Software Int.

Thissen, D., Steinberg, L., and Wainer, H. (1988). Use of item response theory in the study of group differences in trace lines. In H. Wainer and H. I. Braun (Eds.), *Test Validity* (pp. 147–169). Hillsdale, NJ: Lawrence Erlbaum Associates.

Thurstone, L. L. (1931). Measurement of social attitudes. *Journal of Abnormal and Social Psychology, 26,* 249–269.

Urry, V. W. (1974). Approximation to item parameters of mental test models and their use. *Educational and Psychological Measurement, 34,* 253–269.

Vale, C. D. (2006). Computerized item banking. In S. M. Downing and T. M. Haladyna (Eds.), *Handbook of Test Development* (pp. 261–285). Mahwah, NJ: Lawrence Erlbaum Associates.

Van den Noortgate, W., and Onghena, P. (2005). Meta-analysis. In B. S. Everitt and D. C. Howell (Eds.), *Encyclopedia of Statistics in Behavioral Science* (Vol. 3, pp. 1206–1217). Chichester: Wiley.

Van der Linden, W. J., and Boekkooi-Timminga, E. (1989). A maximum model for test design with practical constraints. *Psychometrika, 54,* 237–247.

Van der Linden, W. J., and Glas, C. A. W. (2000). *Computerized Adaptive Testing: Theory and Practice.* Dordrecht: Kluwer Academic.

Van der Linden, W. J., and Hambleton, R. K. (Eds.). (1997). *Handbook of Modern Item Response Theory.* New York: Springer-Verlag.

Van der Linden, W. J., and Mellenbergh, G. J. (1977). Optimal cutting scores using a linear loss function. *Applied Psychological Measurement, 1,* 593–599.

Van der Linden, W. J., and Reese, L. M. (1998). A model for optimal constrained adaptive testing. *Applied Psychological Measurement, 22,* 259–270.

Van der Linden, W. J., and Veldkamp, B. P. (2004). Constraining item exposure in computerized adaptive testing with shadow tests. *Journal of Educational and Behavioral Statistics, 29,* 273–291.

Van der Rijt, B. A. M., Van Luit, J. E. H., and Pennings, A. H. (1999). The construction of the Utrecht Early Mathematical Competence Scales. *Educational and Psychological Measurement, 59,* 289–309.

Verhelst, N. D., and Glas, C. A. W. (1995). The one parameter logistic model. In G. H. Fischer and I. W. Molenaar (Eds.), *Rasch Models: Foundations, Recent Developments and Applications* (pp. 215–237). New York: Springer.

Von Davier, A. A., Holland, P. W., and Thayer, D. T. (2004). *The Kernel Method of Test Equating.* New York: Springer.

Von Davier, A. A., and Kong, N. (2005). A unified approach to linear equating for the nonequivalent groups design. *Journal of Educational and Behavioral Statistics, 30,* 313–342.

Wainer, H., and Kiely, G.L. (1987). Item clusters and computerized adaptive testing: a case for testlets. *Journal of Educational Measurement, 24,* 185–201.

Wainer, H., and Wang, X. (2000). Using a new statistical model for testlets to score TOEFL. *Journal of Educational Measurement, 37,* 203–220.

Waller, N. G. (1998). Review of the Beck Depression Inventory (1993 revised). In J. C. Impara and B. S. Plake (Eds.), *The Thirteenth Mental Measurement Yearbook* (pp. 120–121). Lincoln, NE: The Buros Institute of Mental Measurements.

Wang, T., and Zhang, J. (2006). Optimal partitioning of testing time: theoretical properties and practical implications. *Psychometrika, 71,* 105–120.

Wang, W. -C., and Su, Y. -Y. (2004). Factors influencing the Mantel and generalized Mantel-Haenszel methods for the assessment of differential item functioning in polytomous items. *Applied Psychological Measurement, 28,* 450–480.

Warm, T. A. (1989). Weighted likelihood estimation of ability in item response theory. *Psychometrika, 54,* 427–450.

Webb, N. M., Shavelson, R. J., Kim, K. S., and Chen, Z. (1989). Reliability (generalizability) of job performance measurements: Navy machinist mates. *Military Psychology, 1,* 91–110.

Werts, C. E., Breland, H. M., Grandy, J., and Rock, D. R. (1980). Using longitudinal data to estimate reliability in the presence of correlated measurement models. *Educational and Psychological Measurement, 40,* 19–29.

Wilcox, R. R. (1976). A note on the length and passing score of a mastery test. *Journal of Educational Statistics, 1,* 359–364.

Wilcox, R. R. (1981). A closed sequential procedure for comparing the binomial distribution to a standard. *British Journal of Mathematical and Statistical Psychology, 34,* 238–242.

Wilhelm, O., and Schulze, R. (2002). The relation of speeded and unspeeded reasoning with mental speed. *Intelligence, 30,* 537–554.

Williamson, D. M., Almond, R. G., Mislevy, R. J., and Levy, R. (2006). An application of Bayesian networks in automated scoring of computerized simulation tasks. In D. M. Williamson, I. I. Bejar, and R. J. Mislevy (Eds.), *Automated Scoring of Complex Tasks in Computer-Based Testing* (pp. 201–257). Mahwah, NJ; Lawrence Erlbaum Associates.

Wilson, D. T., Wood, R., and Gibbons, R. (1991). *TESTFACT: Test Scoring, Item Statistics, and Item Factor Analysis.* Mooresville, IN: Scientific Software.

Wise, S. L., and DeMars, C. E. (2006). An application of item response time: the effort-moderated IRT model. *Journal of Educational Measurement, 43,* 19–38.

Wollack, J. A., and Cohen, A. S. (1998). Detection of answer copying with unknown item and trait parameters. *Applied Psychological Measurement, 22,* 144–152.

Woodruff, D. (1990). Conditional standard error of measurement in prediction. *Journal of Educational Measurement, 27*, 191–208.

Wright, B. (1988). The efficacy of unconditional maximum likelihood bias correction: comment on Jansen, van den Wollenberg, and Wierda. *Applied Psychological Measurement, 12*, 315–318.

Wright, B.D., and Stone, M.H. (1979). *Best Test Design.* Chicago: Mesa Press.

Yen, W.M. (1981). Using simulation results to choose a latent trait model. *Applied Psychological Measurement, 5*, 245–262.

Zimowski, M., Muraki, E., Mislevy, R. J., and Bock, R. D. (1996). *BILOG-MG: Multiple-Group IRT Analysis and Test Maintenance for Binary Items.* Chicago: Scientific Software.

Zwick, R. (1990). When do item response function and Mantel-Haenszel definitions of differential item functioning coincide? *Journal of Educational Statistics, 15*, 185–197.

Author Index

Subject Index

Milton Keynes UK
Ingram Content Group UK Ltd.
UKHW040444071024
449327UK00020B/969

9 780367 388676